▶英語で学ぶ自然科学シリーズ **2**
Introductory Series on Natural Sciences

英語で学ぶ環境科学
— 食料, 資源と環境 —

Introduction to Environmental Sciences and Practices
— Food, Natural Resources and Environment —

Ph.D. 渡邉 和男 編著

渡邉 純子 著

コロナ社

「英語で学ぶ環境科学」初版2刷正誤表

Page	Line	Error/Correction	Recommendation
iv	1	How Would be Our	How Would Our
1	5	Twenty fours hours	Twenty four hours
p.1:13行, p.5:1,15行, p.11:18行, p.12:26行, p.15:9行, p.18:4.-(5), p.21:12行, p.25:2.-(1), p.30:3行, p.31:1行, p.33:2.-(3), p.46:3,7,12行, p.50:10行, p.58:11行, p.95:5行, p.147:31行		stuffs	stuff
1	21	.?	. (?トル)
2	5	suffered	suffering
3	2	rice is self-sufficient	rice production
3	7	its survival in term of	its survival in terms of
4	3	unnecessary too high	unnecessarily too high
5	20	one of options	one of the options
6	5	Shall we consumes	Shall we consume
6	8	know that when these	know when these
6	8	at its season.	at its season?
6	13	and proud of them	and be proud of them
8	2.-(2)	Wheat is highly self-sufficient in Japan	Japan is highly self-sufficient in wheat
8	2.-(3)	Japan over-produces rice and export a lot to USA	Japan overproduces rice and exports a lot to the USA
11	5	Japanese depend foreign countries	Japanese depend on foreign countries
12	23	Japan?	Japan.
13	7	gasses	gases
13	9	Cheaper, easier and demands	Cheaper and easier production and large demands
13	14	litters	liters
14	Figure2.2 タイトル2行目	litten	liter
15	6	spends	spend
15	12	in term	in terms
18	4.-(1)	think importing food	think of importing food
18	4.-(2)	What are alternative	What are the alternative
18	4.-(4)	Are Food mileage	Are food mileage
18	4.-(4)	foods	food
19	3	What's else	What else
20	12	noodles etc.	noodles, etc.
21	1	vitamin	vitamins
21	6	fish and eggs.	fish, and eggs.
22	Figure 3.2 の左上	kidny	kidney
26	4.-(3)	do also maintain	do to maintain
28	15	grand children	grandchildren
29	25	felling	feeling
30	1	such at	such as
30	11	take costs	takes costs

Page	Line	Error/Correction	Recommendation
31	6	internets	internet
31	9	the miss-uses	the misuses
31	14	had been suffered form	suffered from
31	16	worried how	worried at how
31	24	suppliers etc?	suppliers, etc?
33	2.-(1)	an health problem	a health problem
35	4.-(5)	food is	food if it is
36	Figure 5.1	Pnicipitation	Precipitation
37	1	Only limited number of species	Only a limited number of species
37	8	cloths	clothes
38	1	313 litters	313 liters
38	3	1,460 litters	1, 460 liters
39	Figure 5.2 の右上	Area of sea area	Area of sea
40	21	wouldn't be selfish	wouldn't we be selfish
41	7	Iraq face the	Iraq are facing
41	21	water is one of fundamental	water is one of the fundamental
46	2	An Average Japanese produce	An average Japanese produces
46	5	The garbage come from kitchen	The garbage coming from the kitchen
49	3	such a large items	large items
50	1	Recycling	recycling
50	15	individuals guaranteed	individuals is guaranteed
52	2.-(5)	dose	do/A Japanese does
53	4.-(1)	it burden	it a burden
54	4.-(5)	once I abandon	once they are abandoned
55	6	one of major reasons	one of the major reasons
58	11	Food stuffs are	Food stuff is
58	12	been taken places already in many places in	been taking places already around
60	9	More	more
60	10	are already suffered constantly from huger	have been suffering constantly from hunger
62	5	ofon	of on
63	4.-(2)	feed, industrial	feed, and industrial
63	4.-(3)	do think engaging	do you think about engaging
65	3	the rare	rare
65	14	similar each	similar to each
66	20	consume the sixty percent	consume sixty percent
66	22	gram of	grams of
69	22	Kazafstan	Kazakhstan
70	17	have own	have its own
75	5	is the introducer of the	makes
75	10	blooming in the spring	blooming in spring
75	22	when a	where a
75	27	which fruits ripen	which the fruit ripens
76	4	He founds	He finds
78	5	an important sources for a life	an important source for life
80	2.-(4)	Crawhfish	Crawfish
81	4.-(1)	is difference	is the difference
83	4	of major	of the major
83	6	over seas	overseas
83	22～25の 1)～4)	, （カンマ）	; （セミコロン）

Page	Line	Error/Correction	Recommendation
84	16	at enormous	at an enormous
84	23	is important	is an important
89	7	a exotic	an exotic
89	9	a logical	logical
89	14	be balance	be a balance
91	7	the humans	humans
93	4.-(4)	species been benefited	species benefited
94	3	Inanimate beings are such as atmosphere,	Inanimate beings are atmosphere,
96	10	degree celcius	degrees celsius
96	16	sadness of	degradation of
96	22	that average	that an average
98	11	activities? We can save water such	activities. We can save water by
99	1	cloths and use rain for flushing toilet	clothes and use rainwater for flushing the toilet
99	5	domestic electric equipments	domestic electric equipment
99	8	should we	we should
99	10	the plastic	plastic
99	12	we to try	we try
102	4.-(1)	form the	from the
105	4	of major	of the major
105	8	the stakeholders	stakeholders
105	13	?	． (ピリオド)
105	15	Anyway	Nevertheless
105	18	Limit	limit
105	24	20 th century	20th century
105	25	for United	for the United
105	26	he become a	he became a
105	28	personal incline to the nature	personal incline to nature
106	3	about a book	about the book
106	14	address	addresses
106	18	lessens	lessons
106	20	With former	With the former
107	11	disturb agriculture	disturbs the agriculture
107	16	biethics	bioethics
108	19	mind four	mind these four
111	2.-(3)	simple	a simple
114	1	How Would be Our	How Would Our
114	3	waste	wastes
117	25	capacity what	capacity of what
125	25	a essential	an essential
127	2	Internaitonal	International
145	28	we depend a lot of food overseas	we depend a lot on food overseas
145	29	even USA	even in the USA
147	4	dietitian	a dietitian
148	17	rain fall	rainfall
150	2	the idiotic act	idiotic act
151	27	sources for	source for
151	28	it totally	it is totally
153	10	gram of	grams of
153	11	gram	grams
153	31	a evergreen	an evergreen
157	5	sustinability	sustainability

Preface

　あらゆる科学技術分野での情報は，電子媒体を主体に容易にアクセスすることができる．一方，専門的な先端的科学技術情報を，正確にかつ迅速に理解することは容易ではない．そして，多くの有用な情報は，英語で大量に流通している．このような背景を考慮して，英語で学びながら特に自然科学分野での情報理解を支援するために，この「英語で学ぶ自然科学シリーズ」は企画された．

　本書は，シリーズ第2作目であり，生活に関わる食料，資源と環境科学分野の基礎知識や，グローバリゼーションが進行する中での人の生活と世界との関わりを紹介している．これらについての用語を英語で提供し，英語での情報検索等への支援教材とすることを目的とした．本書によって，現代の日常生活と科学との関わりの紹介を行うとともに，英字新聞や英文雑誌あるいは専門誌の読解の支援となる情報も提供している．

　本書は，食料，資源と環境科学分野を目指す高校生や大学生のみならず，一般読者が先端情報に容易に接する手助けやきっかけになることを目指したものである．内容は，Unit A から Unit D までの4分野に大別されているが，オムニバス形式で，どの章から読みはじめても理解できるように構成されている．

　Unit A では，生存の基盤である食料について解説している．食料保障，世界的な生産と流通での課題や食品の安全性などについて，話題を提供している．

　Unit B では，昨今急激に世界的な大問題になってきている天然資源を取り扱い，特に水やエネルギーなど人間の基本的生存に関わる資源について話題を提供し，かつ産業資源の窮状を指摘している．

　Unit C では環境のトピックスを取り入れ，生活と環境の関わりを身近な事

例を取り上げ紹介している。

　最後の Unit D では，前の三つの Unit を総括し，日常生活の中，どのように環境の保全と持続的利用を実践すればよいのか，その実例と規範について議論している。

　また，Appendix として，今後，日本で大きな話題として取り上げられる，生物多様性の保全と利用についてのテキストを紹介している。

　本書の内容を理解することによって，幅広く環境と天然資源の世界的な課題についての関心が高まることを期待したい。

2008 年 6 月

編著者　渡邉　和男

Contents

Unit A   Foods and Security

Chapter 1　How Can We Get Our Food? ·· 1
Chapter 2　Food Mileage and Virtual Water ····································· 11
Chapter 3　Food Nutritional Value and Health ································ 19
Chapter 4　Food Safety and Information ·· 28

Unit B   Natural Resources

Chapter 5　Water: Problems and Conflicts ······································ 36
Chapter 6　Waste and Domestic Garbage ·· 46
Chapter 7　Energy Demands and New Sources ······························ 55
Chapter 8　Rare Metals and Conflicts ·· 65

Unit C   Environment

Chapter 9　Food Chain and Biodiversity ·· 75
Chapter 10　Invasive Species and
　　　　　　Crisis in Biodiversity ·· 83
Chapter 11　Facing Environmental Degradation ···························· 94

Unit D   Summary

Chapter 12　Environment and Ethics ··· 104

Chapter 13 How Would be Our Efforts to Daily Life
            to be Environmentally Friendly? ································ 114
Chapter 14 Epilogue: Challenges at
            Tsukuba Science City ···································· 122

Appendix ·· 125
References ··· 139
Answers to Exercises ·· 144
Index ·· 160

Unit A　Foods and Security

Chapter 1　How Can We Get Our Food?

We have plenty of food in Japan. We can buy any items as far as we can pay for them. Automatic beverage vending machines are elsewhere. Convenience stores open twenty fours hours everyday and fast food shops provide quick bites. But do you know where these food materials come from?

　　Think about your ordinary breakfast. Two pieces of toasts, a bit of butter, a glass of orange juice, two eggs, a slice of tomato, a finger of banana and maybe a cup of coffee for almost everyday. Of course, you afford them because you want them for your diet and for your preference to maintain your healthy life. But don't you think it is strange that you get such food stuffs year round if you consider seasons of crops, production locations, food safety and quality against so much demands with a large number of Japanese population.

　　We do not have much local production in Japan on many food items. Some crops are seasonal unless we use greenhouses with artificial energy uses. We know that tomatoes do not grow in winter at many locations without supplementary heating. We are spoiling energy and exhausting more carbon dioxide for unintentional purchase of such seasonless crop availability. ?

　　Coffee? The best coffees are brought from all over the world for us. Local people in Columbia can hardly get a high quality coffee at a reasonable price as Japanese and other buyers put the price higher than the local people can afford. Even we cannot grow warm or tropical zone crops sufficiently and efficiently in Japan. Banana is fully imported from tropical countries where no synthetic energy sources are required.

We must know that a lot of agrochemicals are used for maintaining production and quality for the consumers who demand nice looking such as uniformity as well as safety, and farmers overseas are often suffered from overdoses such by pesticides and fungicides directly and also from the polluted environments as the consequences. Do we know that?

While we can grow wheat for breads and noodles at many localities in Japan, we do not have much supply which meet with the present demands, and consequently a lot of imports are made and average ninety percent of demands are from overseas in the case of wheat. Only forty percent of the Japanese food demand is produced domestically, and the rest is imported globally (**Figure 1.1**).

It should also be pointed out that only twenty-four percent of

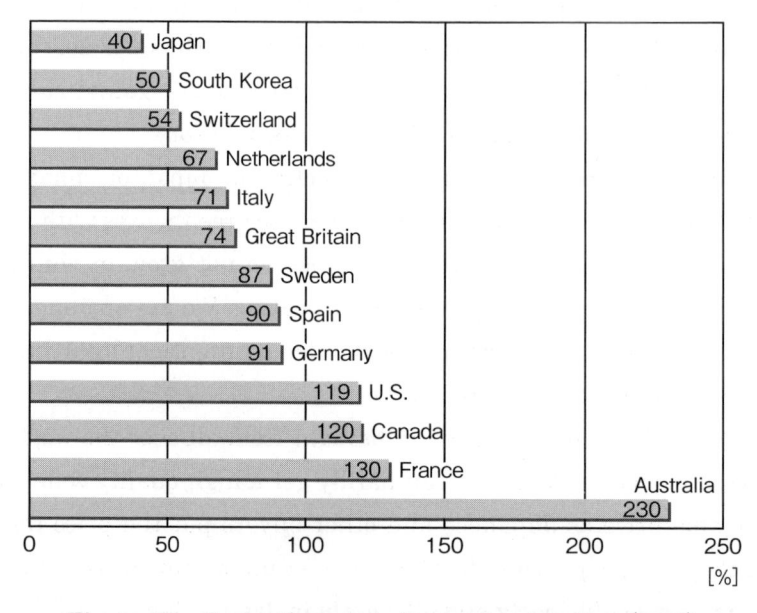

Country	%
Japan	40
South Korea	50
Switzerland	54
Netherlands	67
Italy	71
Great Britain	74
Sweden	87
Spain	90
Germany	91
U.S.	119
Canada	120
France	130
Australia	230

Figure 1.1 Food self-supply in the developed nations (2002) (modified from FAOSTAT)

Chapter 1   How Can We Get Our Food?   **3**

demanded grains and pulses are self-supplied in Japan (**Figure 1.2**): rice is self-sufficient, but not on wheat, barley, soybean etc. Furthermore, many developed countries such as USA and EU nations, are self-sufficient of food supplies and even they are exporters of specific food items.

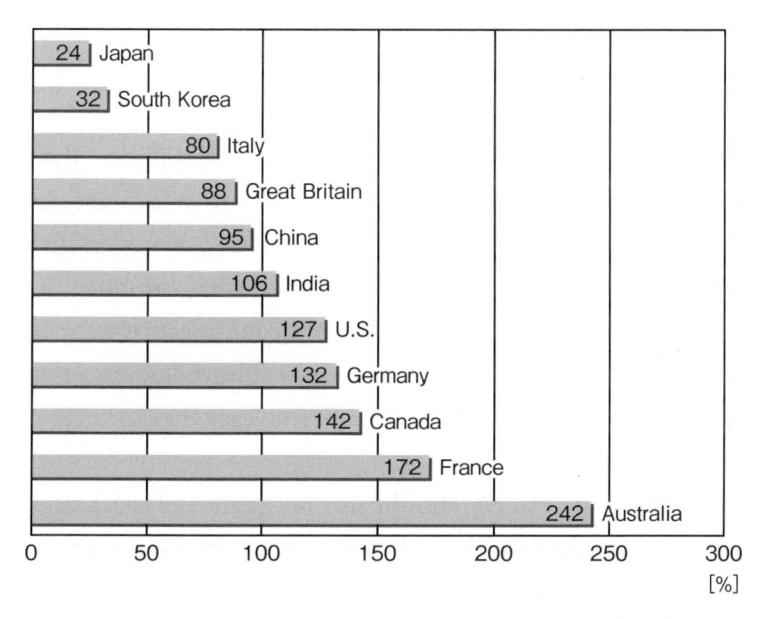

Country	%
Japan	24
South Korea	32
Italy	80
Great Britain	88
China	95
India	106
U.S.	127
Germany	132
Canada	142
France	172
Australia	242

Figure 1.2   Grain self-supply in several countries (2004) (modified from FAOSTAT)

Japan may be handicapped as a nation not to be independent on its survival in term of food supply for citizens. Nowadays, the prices of imports become skyrocketing: the wheat price go up 30 % more at the first quarter of 2008, and Japanese are exposed to the high price and also international competition of supplies. Soon we may dream the udon noodles, Italian pasta and bread in the night as self-production of wheat is very low in Japan.

Also price and quality are of concerns. With cheaper labor and production costs at foreign countries, although the aliments are transported through a long distance, the end-retail prices in Japan are cheaper than how Japanese local producers provide. The demands are unnecessary too high from Japanese consumers, optimum hygiene and safety should be guaranteed accordingly to the standards and accreditations given by the buyers, importing nations and consequently by the consumers! However, do we need exactly the same size and shape of vegetables such as asparagus that are often from Southeast Asia and Latin America.

This makes more difficulty for producers and also on the value competitiveness. It also should be addressed that we are not the only one nation buying food from those countries, now many Asian countries afford foreign food material with increasing number of more wealthy consumers, and they have also strong incentive for having more and better food materials. Japanese will have less opportunities to have the required quantity and quality at ease due to such emerging demands from other countries.

The domestic efforts should not be neglected. Although less population of farmers dedicate to agriculture in Japan, they make very serious efforts to meet with the needs of the folks for the price and quality.

Also there should be consideration and more debating on the subsidies. Food security and the protection of farmer communities are another aspects at the national and also local level, respectively. Encouragement in competitiveness is easily pointed out, however, is there integrated system made to back-up such initiation of the effort?

We examined the consumer views of economics and availability. Then let's talk about the national policy and international relationships, Why do we import food? The consumer need is not only one reason but also international trade balance is another. Government encourages the trade-off between high value industrial products such as cars and electric equipments and the food stuffs. It is always debatable on how to protect Japanese domestic agriculture against the elements at WTO negotiations, and on the cumbersome local politics on agriculture sector. Consumers are the one of the stakeholders on the issues, and should we participate more actively on what we can eat every day?

You may feel inconvenience if you cannot get tomatoes in winter at Tokyo precincts. You would be depressed if you have less option to buy from supermarkets as apples are out of season in summer. Then, mentally you may have difficulty. But think if a bit of patience makes your environment better, if overall price changes to lower, and if the world would be more sustainable in a long term. Indeed, we may not need so much of imports.

A large portion, which is around forty percent of imported food stuffs are spoiled. Also many Japanese concern obesity and metabolic syndromes. Under this fact, do we really need keeping enormous importation? It shall help wasting by a small effort of saving the uses and refraining the purchases?

Should we think and practice the concept on Local Production and Local Consumption? This is one of options for promoting local food production toward increasing more chance of agriculture revival, subsequently being a small component of alleviating national food security issues. Also it is logical that the fresh products at season have high nutritional values and tastes than artificially made such as greenhouses.

There are needs for taking into account on the balances of price,

Notes :
WTO：World Trade Organization 「世界貿易機関」
Local Production and Local Consumption 「地産地消」

energy uses, safety controls, labor opportunities etc. This also increases rural community rejuvenation by the promotion of the old generations to be active and overall labor and income opportunity, local landscape conservation and sustainable uses etc.

Shall we consumes seasonal and locally available items such as cucumbers, eggplants and tomatoes in summer and Chinese cabbage, white radish and apples in cool period as much as possible? Did you know that when these crops are at its season. Also it is true that consumers shall contemplate on discouragement of buying the imported and artificial energy-invested local materials which are supplied against off-seasons such as tomatoes.

We shall promote the products made locally at season in Japan and proud of them.

Exercises

1. つぎの1〜71の語に対応する英単語または英熟語を本文から選び出して書き込み，また発音しなさい（動詞は原形を記入しなさい）。

	Japanese	English		Japanese	English
1	たくさんの		9	嗜好	
2	支払う		10	補足の	
3	購う		11	むだにする	
4	飲料		12	排出する	
5	売る		13	配慮のない	
6	通常の		14	十分な	
7	購入する		15	効率的な	
8	常食		16	合成の	

Chapter 1　How Can We Get Our Food?

	Japanese	English		Japanese	English
17	輸入		38	衛生	
18	農薬		39	認定基準	
19	維持する		40	民衆	
20	過剰投与		41	無視する	
21	殺虫剤		42	献身する	
22	殺菌剤		43	議論する	
23	汚染する		44	勧める	
24	結果		45	競争力	
25	害を受ける		46	統合する	
26	供給		47	開始	
27	需要		48	検分する	
28	国内の		49	政策	
29	穀類		50	取引/駆け引き	
30	豆類		51	やっかいな	
31	負担のある		52	不便	
32	国家		53	交渉	
33	生存		54	分野/業界	
34	食料		55	関係者	
35	移送		56	近郊	
36	小売り		57	肥満	
37	価格		58	メタボリック症候群	

8   Unit A   Foods and Security

	Japanese	English		Japanese	English
59	復興		66	収入	
60	再生		67	労力	
61	勧奨/促進		68	勧めない	
62	景観		69	熟考する	
63	保全		70	投資	
64	考慮する		71	誇りである	
65	田舎の				

2．つぎの各文が本文の内容と一致するものにはT(True)，一致しないものにはF(False)を，文末の（　）に記入しなさい。
（1）Japanese do not depend heavily on the imported food.（　）
（2）Wheat is highly self-sufficient in Japan.（　）
（3）Japan over-produces rice and export a lot to USA.（　）
（4）Japanese food quality requirement is high and we often pay for the price.（　）
（5）Many EU countries have high rates of self-supply of aliments.（　）

3．つぎの日本語の各文を（　）の中の語を用いて英語の文にしなさい（必要があれば単語を適切な形に変換しなさい）。
（1）朝市に行けば，季節の新鮮な野菜がたくさん買える。(plenty of, purchase)

（2）季節はずれの野菜や果物を供給するために，われわれは大量のエネルギーを浪費し，環境を汚染している。(supply, waste, pollute)

（3）日本の農業の国際競争力は，人々によって頻繁に議論されるべきである。(debate, competitiveness, folk)

（4）生ゴミを減らすことを勧めましょう。(promote, garbage)

（5）一人ひとりの日本人が，食料保障について真剣に考えるべきである。(contemplate, food security)

4．つぎの各問いに英語で答えなさい。
（1）Should we reduce food import?

（2）Can Japan produce more food?

(3) Is it discouraged to buy imported food?

(4) Is national food security more important by increasing self-supply or international diplomacy on trade?

(5) How can you avoid food spoilage and waste?

5. つぎの日本語課題を英語で議論してください。
（1）日本の食料自給と海外諸国の食料事情とを比較して，日本での食料生産や農業支援はどうあるべきか？

（2）世界的に食料の価格が高騰している。燃料輸送費も高騰し続けてきている。地球の気候変動に農業は多大な影響を受け，水の供給も大きな課題になってきている。世界的に食料保障をどのように考え対処すべきであろうか？

Chapter 2   Food Mileage and Virtual Water

As in the Chapter 1, we had discussion on food security and debates on domestic production. Here we will expose ourselves further on the food importation.

Japanese depend foreign countries on our food. Majority of our demands on soybean, corn, and wheat are brought from many nations. Canola oil is almost one hundred percent imported. Even those aliments travel a long distance from other side of this planet. Many of ordinary fresh food such as fish, fruits and vegetables are flying trans-oceans. For example, your fried shrimp, ebi-furai, flies over five thousand kilometers from Southeast Asia such from Thailand.

Transport is one to concern and also choice is another to think about. Japanese consumers demand the safety issues to the suppliers, but we have to recognize that those farmers and dealers produce and handle for their business with maximum profits for their living. Thus, labor and cost efficiencies are important besides assuring the safe quality of the food stuffs for sales.

The agricultural and other food products are generally regarded as safe as far as produced and handled with common knowledge, accredited or standardized procedures such as HACCP, and guidance given by public sector, however, sometimes Japanese consumers are not happy yet on the imports and require producers overseas on the

Notes :
Canola oil「不飽和脂肪酸の比率の高い菜種油」
HACCP : Hazard Analysis and Critical Control Point「食品の衛生管理の手法，ハサップ」

choices of crop varieties, production methods, handlings and transports. Japanese are demanding more than how we shall be?

　Fish are delivered also from all over the world. Eels and shrimps from China and Southeast Asia, clams from Korea, mackerel and salmon are from North Sea especially from Norway, crabs from Russia, octopus from Africa and South America and our favorite tunas from all oceans. Often Japanese fishermen catch the needs at distant oceans, but since those catches do not meet enough with the market demands in Japan, imports are also made in a large quantity from those regions in the world.

　Chicken was used to be produced, processed and delivered from Southeast Asia, but due to increasing concerns on pandemic possibility of zoonosis on avian influenza in Asia, the suppliers have been diverted to far long distant locations such as Brazil. It is over twenty thousand kilometers from Brazil by traveling over half of the round of the planet. Beef is also brought from USA, Australia and South American countries. All locations are very distant.

　It is not a long history that suppliers had troubles to provide quantity of beef to fast-food chains and tables of individual families without concerns of BSE in USA, and we realized that our food supply system is very fragile. Alternative ways should be well planned and we must be ready if food-supplying nations were to have a trouble in production or stopping exports to Japan?

　Why do we import in such a way? Trade matters, availability in amount and at season, costs and even safety are the reasons. Because of such a long distance transportation of food stuffs, Japanese exhaust

Notes :
　avian influenza「トリインフルエンザ」
　BSE : Bovine Spongiform Encephalopathy「牛海綿状脳症, 狂牛病」

three times more carbon dioxide than Koreans do on aliment purchase from overseas, and nine times more compared with French (**Figure 2.1**). Yet, those sacrifices on energy and emission of carbon dioxide, trade off the needs and out-beat the domestic production.

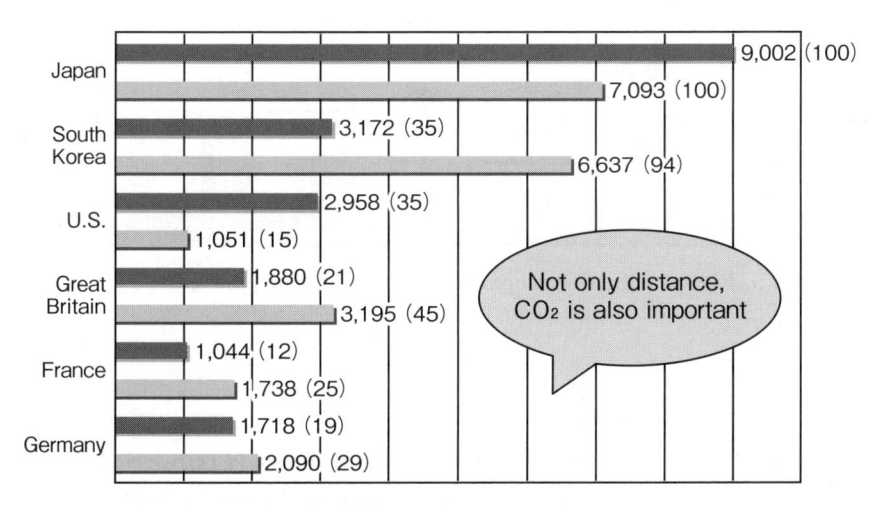

```
Japan        9,002 (100)
             7,093 (100)
South Korea  3,172 (35)
             6,637 (94)
U.S.         2,958 (35)
             1,051 (15)
Great Britain 1,880 (21)
             3,195 (45)
France       1,044 (12)
             1,738 (25)
Germany      1,718 (19)
             2,090 (29)
```

Not only distance, CO_2 is also important

Figure 2.1 The comparison of food miles in the world
(Modified from http://www.e-shokuiku.com/jyukyu/13_3.html and http://www.eic.or.jp/ecoterm/?act=view&serial=2759)

Food mileage comes from the fact of such a long distance transport of the food materials, often such transport actions cause the energy waste and over-exhaust of the global warming gasses. It is required for sixteen thousand kilometers to assemble a lunch box, or a bento at a convenience store in Japan. Cheaper, easier and demands are reasons to buy the materials overseas, however, should we consider on the consequence at a long term.

Food production is not only on energy and emission, but also on water uses (**Figure 2.2**). Do we know that one kilogram of brown rice is the result of the use of 2,000 to 4,500 litters of fresh water over the

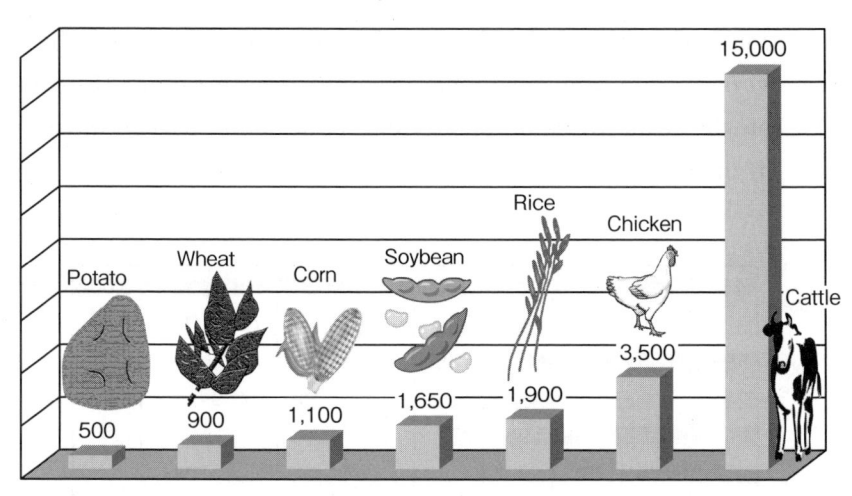

Figure 2.2 The comparison of water for food production (litten for 1 kg of food)

Potato 500, Wheat 900, Corn 1,100, Soybean 1,650, Rice 1,900, Chicken 3,500, Cattle 15,000

crop production period? When field crops such as corn, soybean and wheat are grown, yet, average 2,000 tons of water is required for one ton of the harvest. A beef rice bowl, gyudon, is the result of the use of 2 tons of water for feeding beef. It ten times more than an average amount of water in a Japanese bath.

We ask foreign countries to produce food, and actually we consume their water, too. This is called virtual water indication, how the food production requires water. Japanese consume the total of eight seven billion ton per year for daily and industrial activities, and in addition to them, virtual water consumption is sixty four billion tons per year as we import a lot of food.

The world needs more fresh and clean water for survival of all human beings. Many people at developing countries face to health problems due to the lack of quantity and quality of clean fresh water. Also they need water for food production. Even in United States of America, where they produce maize, soybean and other export commodities, they are short of waters both for urban and agricultural uses.

Furthermore, drought is the inherent headache for agriculture and conservation of vegetation in many Southern and Western parts of USA, and water trade among states are very critical issues. Water is essential for all, and Japanese spends enormous amount of water overseas by buying the products over there.

Should we realize that we may waste energy and pollute more environments by importing food stuffs and other materials overseas which we may be able to produce domestically with an efficiency. Debates also should be encouraged whether domestic products are superior to imports in term of such environmental concerns, cost, political aspects, trade and social issues. Also we must always take into account that the production of aliments, particularly agricultural entities, requires a lot of water, especially precious fresh water which is essential for human survival.

Exercises

1. つぎの1〜40の語に対応する英単語または英熟語を本文から選び出して書き込み，また発音しなさい（動詞は原形を記入しなさい）。

	Japanese	English		Japanese	English
1	議論		6	確信させる	
2	大洋を越えて		7	基準を認める	
3	認知する		8	標準化する	
4	取扱い業者		9	指導	
5	利益		10	ウナギ	

16 Unit A Foods and Security

	Japanese	English		Japanese	English
11	アサリ		26	食料	
12	サバ		27	排気する	
13	サケ		28	排出	
14	カニ		29	犠牲	
15	タコ		30	駆け引きする	
16	マグロ		31	打ち負かす	
17	配送する		32	実質上の	
18	流行性の		33	指標	
19	人畜共通感染症		34	消費する	
20	転換する		35	消費	
21	脆弱な		36	生来の	
22	他の		37	頭痛	
23	輸入		38	商品	
24	輸出		39	存在	
25	取引		40	不可欠な	

2. つぎの各文が本文の内容と一致するものにはT(True)，一致しないものにはF(False)を，文末の（　）に記入しなさい．

(1) We do not import fish from overseas. (　　)

(2) Edible oils are produced domestically with a high sufficiency in Japan. (　　)

(3) We do not consume a lot of water overseas by demanding the farmers over there on our food. (　　)

(4) Importing food also makes the emission of carbon dioxide due to the transportation. (　　)

（5）Bentos sold at convenience stores, are the result of the assemblage of the food transported from long distances. (　　)

3. つぎの日本語の各文を（　　）の中の語を用いて英語の文にしなさい（必要があれば単語を適切な形に変換しなさい）。

（1）日本人は世界中から食料を輸入しているだけではなく，エネルギーや水も同時に消費している。(consume, aliments)

（2）トリインフルエンザのヒトへの流行性感染の可能性を恐れて，日本人は遠隔地のブラジルから大量に鶏肉を輸入している。(avian influenza, pandemic breakout)

（3）食料を大量に輸入している限りは，日本の食料保障は脆弱である。(depend, fragile, food security)

（4）食料輸出国の米国でも，干ばつや水不足は深刻である。(shortage, drought)

Unit A　Foods and Security

（5）われわれは，環境を犠牲にして食糧生産や移動するために二酸化炭素を排出している。(emission, sacrifice)

4．つぎの各問いに英語で答えなさい。

（1）How do you think importing food?

（2）What are alternative options to get fish and aquatic food materials instead of purchasing the imports?

（3）How do you challenge your life with less water uses?

（4）Are Food mileage and virtual water index are absolute entities for decision making on importing foods?

（5）Is the world less interesting if we restrict the trade of food stuffs and promoting domestic products?

Chapter 3   Food Nutritional Value and Health

How do we know that the food we are eating, has relevant nutritional values for keeping us healthy? What's else do we need intakes for supplements for the aliments? How can we distinguish different categories of the food: dietary supplement, designated nutraceutical food, functional food and special dietary food? How do we keep ourselves out from the illness? How much exercise should we do also maintain health? In the below, the texts are quoted from the National Institute of Health and Nutrition under the Ministry of Health, Welfare and Labor (http://www.nih.go.jp/eiken/english/research/program_education.html)

"Shokuiku" promotion is for controlling childhood obesity, as well as on diets of the middle-aged/elderly. Working with the nutrition specialists in the clinical/education fields, the nation-wide research is also conducted, aiming to develop the effective nutritional education methods and also to prove their impacts. In addition, the researches on methods of training and information dissemination for the registered dietitians who are actively involved in "Shokuiku" are implemented.

For the adults, it is also important on the daily dietary care and management. There is no universal recommendation as life style and

Notes :
dietary supplement「健康補助食品」
designated nutraceutical food「特定保健用食品の対応語」
functional food「栄養機能食品」
special dietary food「特別用途食品」
Shokuiku「食育」

individual preference change person by person. But researches shall be conducted for Japanese population on the effective methods and evaluation of the dietary care and management system for an early detection of the undernourished elderly including at risk groups, and also for providing the tailor-made services. Based on such research feedback, we shall also expose ourselves for the knowledge for daily food consumption. With such credible information, we also shall be careful on the fraud or misinformation associated with the food uses as further discussed in Chapter 4.

Each of us shall design the nutritional pyramid accordingly to own life style and preferences (**Figure 3.1**). The foundation is carbohydrates such as rice, bread, noodles etc. The materials for immediate energy for daily activities. The second layer is vegetables and fruit

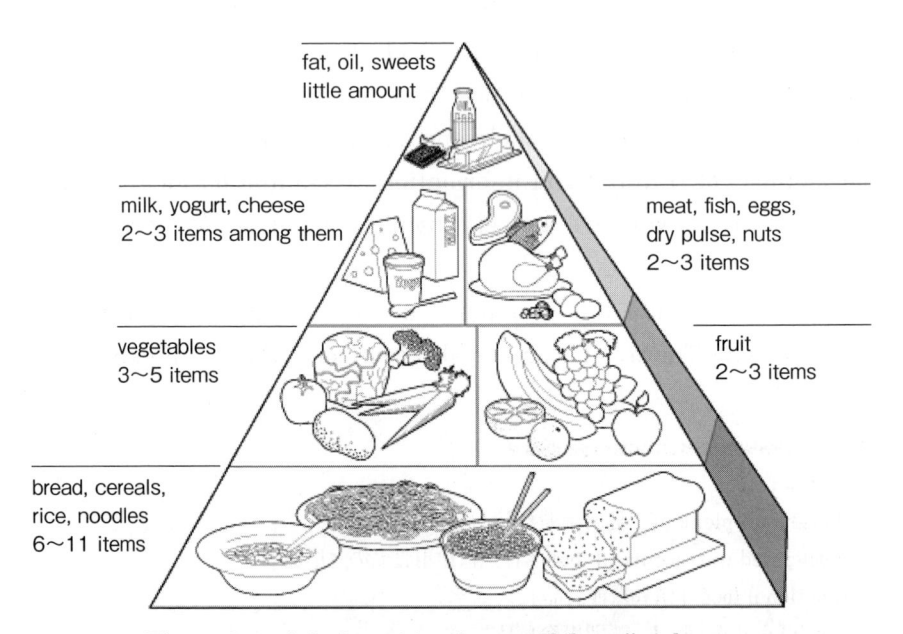

Figure 3.1 A food nutritional pyramid (http://mmh.banyu.co.jp/mmhe2j/sec12/ch152/ch152g.html)

with a bit of seaweed and mushrooms. The category provides vitamin, fibers and minerals. Many Japanese nutritional scientists and doctors, recommend take more vegetables than fruit as fruit often contain excess sugars that can be nurtured at the foundation of carbohydrates.

The third sector is daily products such as milk, cheese and yogurt, and pulses, meats, fish and eggs. They are protein sources. The top layer is oil and sweets with small amount. Overall, we shall know what is the contribution of each food item when they are cooked and eaten (**Figure 3.2**).

The supplementary items such as vitamin or iron tablets, sometimes called "sapuli" in Japan, help only when proper diets are made via food. Those supplementary stuffs cannot replace food and heavy dependency of the complementation could cause physiological disorders and diseases in a long term.

It is also important to know lifestyle-related diseases associated with diets and potential preventions. Among the causes of lifestyle-related diseases, obesity is the serious subject for the affluent Japan, and extensive research undertakes the intervention studies to control/prevent obesity, using the combined behavioral approaches of exercise and diet.

Research and extension are conducted on effective nutritional screening and assessment methods at the dietary care and management, as well as on the scientific procedures and evaluations for planning the tailor-made services for nutrition improvement. Obesity is a primary cause of the lifestyle-related diseases. This effort undertakes intervention studies, where cohort groups of obese subjects are investigated on the effects of exercise and diet for controlling obesity. In addition, a follow-up study will be implemented to examine how long the effects would be sustained after the interventions, and what factors would determine the degree of sustained effects, focusing

22 Unit A Foods and Security

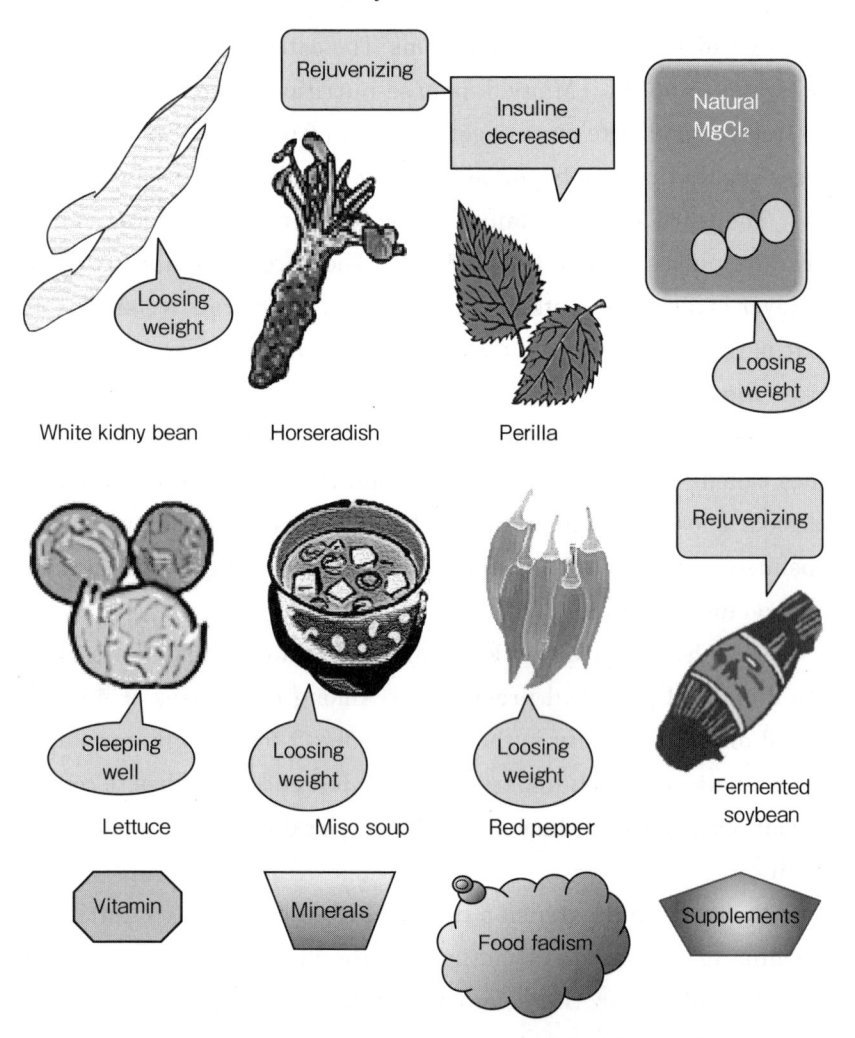

White kidney bean — Loosing weight	Horseradish — Rejuvenizing	Perilla — Insuline decreased	Natural MgCl₂ — Loosing weight
Lettuce — Sleeping well	Miso soup — Loosing weight	Red pepper — Loosing weight	Fermented soybean — Rejuvenizing

Vitamin　　Minerals　　Food fadism　　Supplements

Food fadism is a typical and temporal trend on the food information which may confuse consumers.

Figure 3.2　Many information on foods to enhance our health

Note :
Food fadism「食品に関する一時的な流行」

on the genetic factors.

In response to enactment of "Basic act on Shokuiku", the national campaign is going to be implemented with respect to the health promotion through lifetime efforts for appropriate diets. National project aims to establish the foundation of scientific evidences useful to the promotion of "Shokuiku". And, the researches on childhood obesity, which is closely associated with adult obesity, are undertaken to investigate the effects of nutritional education.

In addition, this project undertakes the researches on the methods of training and information dissemination for the registered dietitians who are actively involved in "Shokuiku" at regional level.

With respect to the National Health and Nutrition Survey, in addition to the response to newly reported food items/menu and the Dietary Referenced Intakes (DRIs) revised every 5 years, the technical and academic investigations are made to improve the accuracy of data management. For which, we work on the routine data management, as well as the researches to enhance the functions of related surveillances.

Exercises

1. つぎの 1 ～ 48 の語に対応する英単語または英熟語を本文から選び出して書き込み，また発音しなさい（動詞は原形を記入しなさい）。

	Japanese	English		Japanese	English
1	適切な		2	栄養の	

Notes :
National Health and Nutrition Survey「国民栄養調査」
Dietary Referenced Intakes (DRIs)「食事摂取基準」

24 Unit A Foods and Security

	Japanese	English		Japanese	English
3	値		25	無機物	
4	摂取		26	過剰の	
5	食物		27	養育する/滋養	
6	補足		28	基盤	
7	引用する		29	豆類	
8	振興		30	効用/貢献	
9	肥満		31	置換する	
10	食事		32	補足	
11	提供する		33	生理的な	
12	登録する		34	異常	
13	栄養士		35	予防	
14	実行する		36	富裕な	
15	世話		37	広範な	
16	個別に設定した		38	阻害	
17	露出する		39	行動の	
18	消費		40	普及活動	
19	信用できる		41	評価	
20	虚偽		42	評価	
21	不適正な情報		43	適正な	
22	栄養摂取ピラミッド		44	調査	
23	好み		45	改訂する	
24	繊維		46	恒常的な	

	Japanese	English		Japanese	English
47	促進する		48	調査	

2. つぎの各文が本文の内容と一致するものにはT(True)，一致しないものにはF(False)を，文末の（　）に記入しなさい。

（1）We can eat a lot of sweet stuffs daily. (　　)
（2）Everybody should have the same diet style. (　　)
（3）Obesity is not a serious health problem in Japan. (　　)
（4）Each of us should have own judgment for food preferences. (　　)
（5）Food values may change as time goes by. (　　)

3. つぎの日本語の各文を（　　）の中の語を用いて英語の文にしなさい（必要があれば単語を適切な形に変換しなさい）。

（1）適切な食事は健康の維持に役立つ。(maintain, health, diet)

（2）肥満は，食事や運動量だけではなく，遺伝的要因にも関係する。(obesity, attribute to, genetic element)

（3）炭水化物は栄養摂取ピラミッドの基盤である。(carbohydrate, foundation, nutritional pyramid)

（４）生活習慣の改善で，病害は予防できる。(prevent, behavioral, improve)

（５）補助食品は通常の食材に置き換わるものではない。長期的な過剰な依存は生理障害や疾病のもとになり得る。

4. つぎの各問いに英語で答えなさい。

（１）How do we know that the food we are eating, has relevant nutritional values for keeping us healthy?

（２）What's else do we need taking for supplements for the aliments?

（３）How much exercise should we do also maintain health?

（４）How do we keep ourselves out from the illness?

(5) Is an education important for food diet?

5．つぎの日本語課題を英語で議論してください．
（1）日本では，大量の食料を輸入している一方，たくさんの食品がむだに捨てられている．日本人一人当たり，毎日 700 カロリーの食料がむだになっており，2,000 カロリーを一人に供給したとして，これは毎日約 4,000 万人の人口を支援できる計算になる．このような現状に個人がどう対処すべきか？

（2）食事は健康の基盤である．一方，知識はあるものの日常生活での実践では，なかなか対処できない．サプリメントなどの利用も考えられるが，自己の生活習慣を見直してみよう．

Chapter 4 Food Safety and Information

Consumers concern many aspects of food. Those worries can be: nutritional values, allergic substances, chemical residues from production such as pesticides, contaminations by pollution such as heavy metals, food poisons, infectious pathogens such as avian influenza etc. Domestic regulatory systems, industrial efforts and international harmonization are vital for assuring the safety and relief to the public.

When you buy vegetables from a local farmers' market with organic products, you may see a green caterpillar in the cabbage, which bites the leaves. How do you feel? It is good that insect like eating the veggie, so that you feel that you have no problem with pesticides and you can eat the cabbage with safety. Or some may be astonished with their imaginary belief that the insect may be parasitic and dangerous, and the cabbage should not be eaten. Life experience or traditional knowledge such from grand parents to their grand children used to help understanding on such a situation. Usually, such herbivore insects are harmless to human and even it is nutritious with protein.

On the other hand, there are debates that the plants bitten by such insects may produce phytoalexins which is induced by attacks. The chemicals may cause carcinogenic. So that bug control by insecticides may be better than organic products with insect damages.

To meet with dealers and consumers demands, often those horticultural products are managed with agricultural chemicals to avoid crop infestation by insects and pathogens. Indeed, it is helpful to

---
Note :
phytoalexin「ファイトアレキシン」

control certain insects that over-feed the plants, which ends up with no crop or unedible quality on the harvest.

Also some plant pathogens are deleterious to crops, so that systematic applications of fungicides and other chemicals are supporting to avoid excess loss of crops. Principally, modern pesticides, fungicides, herbicides and other chemicals have been well established and the safety has been the major issue on the R&D before the commercial uses elsewhere.

In addition, the applications of such agricultural chemicals are well regulated during production and residues are monitored very carefully by the dealers in Japan and other developed countries. It is also same or similar on the meat, fish and any other aliments which are produced, processed and transported from localities in Japan and abroad for Japanese consumers.

BSE, avian influenza, foot-and-mouth disease, have been concerns for the global citizens, and the potential risks have been well dealt by production, supply and processing chains, and efforts have been also made to alleviate the psychological worries on the imports and also on local products.

There are scientifically-reasonable processes and also are questionable parts, however. Scientific facts do not always correspond with the feelings for the safety. This cause indeed big debates and complex systems have been set up, and tedious and costly implementation of safety assurance are employed to the end-consumers to aim for hundred percent felling safe for them.

Parallel to the domestic regulations, there are international

Notes :
R&D：research and development「研究開発」
foot-and-mouth disease「口蹄疫」

standardization discussed such at Codex Alimentarius Commission and SPS at WTO on the safety issues to harmonize the quality of traded food stuffs worldwide. These discussion and rules are vital to Japan as we imports enormous amount of food from many places in the world.

Endeavors are also made for getting the quality from overseas. Japanese industry does not demand only the safety but also assist the foreign producers to meet with the standards required in Japan. Also frequent examination and inspection are made on such quality guarantee and safety. Of course, these activities are not free of cost and indeed, such quality assurance take costs.

Strong voices of consumers always come to the quality and safety of products. Also imaginary fears and concerns come up on modern technologies such as food additives and biotechnology-derived food. GM crops are typically concerned by such campaigns with no scientific evidence.

There are no accidents nor scientific risk reported on the approved GM crops and products there of in the world, however, a certain civilian groups really concerns on the entities as it is artificially genetic-engineered products. Because of this, labeling under Japanese Agricultural Standard Law (JAS), is compulsory to assure the information delivery to guarantee the right to know and consumers' choice

Notes:

Codex Alimentarius Commission「コーデックス委員会，食品の安全性についての国際基準を協議する組織」

SPS at WTO: Sanitary and Phytosanitary Measures at World Trade Organization「WTO おける SPS 協定（衛生植物検疫措置の適用に関する協定）」

Japanese Agricultural Standard Law「農林物資の規格化及び品質表示の適正化に関する法律」

on the GM stuffs.

The regulators and industries are seriously making efforts to serve commodities to the consumers. JAS also supports general standards of the quality. However, information flow sometimes become barriers for understanding and acceptance of the products. Mass media such as TV and internets have been broadcasting sensational news, incorrect interpretation and even falsified scandals etc.

Accidents also have been come up from such incorrect information on the miss-uses of food materials. For example, with a long history of human life, there is enough understanding and scientific supports that pulses such as soybean and French bean should be cooked and the raw materials should not be consumed as it causes diarrhea. A diet promoting TV program abused the information and audience had been suffered form the program. Food faddism has been making promotion of health but some part sensational and biased. Folks are confusing and worried how they can take reliable information for living. Is there any way to have trustable information? Do we have to be wiser by actively studying the present world such as food issues?

This issue shall be tendered for more open discussion for seeking for more sophisticated consensus on the system for knowing information and feeling safety with trust building ways by all stakeholders. Is there any alleviation by consumers by having own participation, instead of depending on the regulators, producers, suppliers etc?

Note :

JAS : Japanese Agricultural Standard 「日本農林規格」

Unit A　Foods and Security

Exercises

1. つぎの1〜46の語に対応する英単語または英熟語を本文から選び出して書き込み，また発音しなさい（動詞は原形を記入しなさい）。

	Japanese	English		Japanese	English
1	有機農業の		19	殺菌剤	
2	芋虫/毛虫		20	除草剤	
3	咬む		21	危険な	
4	野菜		22	規制する	
5	殺虫剤		23	残留物	
6	驚く		24	食物	
7	想像の		25	地域	
8	寄生の		26	おそれ	
9	草食性の		27	心理的な	
10	無害の		28	手間のかかる	
11	栄養がある		29	標準化	
12	議論（する）		30	調和させる	
13	発がん性の		31	努力	
14	虫		32	試験	
15	園芸		33	検査	
16	感染・汚染		34	保証	
17	病源		35	杞憂	
18	食べられない		36	遺伝子工学による	

Chapter 4　Food Safety and Information

	Japanese	English		Japanese	English
37	強制的な		42	下痢	
38	物品		43	豆類	
39	放送		44	聴衆	
40	扇情的な		45	信頼できる	
41	偽証する		46	信用できる	

2. つぎの各文が本文の内容と一致するものにはT(True)，一致しないものにはF(False)を，文末の（　）に記入しなさい。

（1）Uncooked pulses do not cause an health problem. (　)
（2）Public media are totally trustable. (　)
（3）The use of information and food stuffs are your responsibility, too. (　)
（4）Do we have to be wiser by actively studying the present world such as food issues? (　)
（5）Consumers always come to the quality and safety of products. (　)

3. つぎの日本語の各文を（　）の中の語を用いて英語の文にしなさい（必要があれば単語を適切な形に変換しなさい）。

（1）新技術で作られた食品添加物やバイオテクノロジー由来の産物には，杞憂や懸念が起こっている。(imaginary fear, concern, come up)

（2）消費者は，食品加工業者が内容物の情報を偽装したことに驚愕した。(astonish, falsify, content)

Unit A Foods and Security

（3） 虫に食べられた野菜は食用できないわけではない。(bug, bite, unedible)

（4） われわれはどこから食料を輸入してるか理解しなければならない。(import, food stuff, ought to)

（5） 不適切な報道により視聴者は混乱した。(confuse, broadcast)

4. つぎの各問いに英語で答えなさい。
（1） How do we trust mass media on the food information?

（2） Where can I get trustable information on the safety of food?

（3） Is it ethical that TV programs deliver sensational news on food issues?

（4）How can we make understanding among different thoughts of consumers on food safety ?

（5）Who judge your food is safe to consume?

5. つぎの日本語課題を英語で議論してください。

（1）食品の情報の適正は，法律や製造者責任によって担保されている。受け身で，情報を得ることだけではなく，日々の生活でどのように積極的な理解をもつことができるのでしょうか？

（2）Hazard, danger, risk の違いを認知していますか？

（3）安全とはどのような意味なのか？　安心と違うことは認知できているでしょうか？

Unit B   Natural Resources

Chapter 5   Water: Problems and Conflicts

Any living organisms in our planet need water for their existence. The global demands on food and energy have been always concerns especially at rapidly developing regions such as China and India. Not only for survival, but also for the betterment of life, use of natural resources are essential. Water is one of key natural resources for the survival and continuity of humans and the planet.

How much fresh water we have in this planet? We have 1.4 billion cubic kilometer of water in the earth. The majority of the water in our planet is brine or sea water from oceans (**Figure 5.1**).

Many upland living creatures cannot live with saline water.

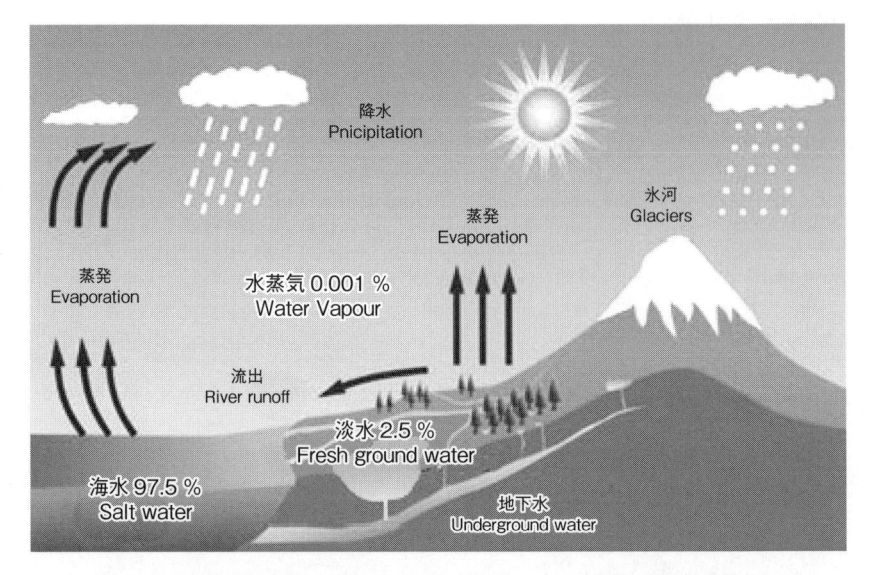

Figure 5.1   Freshwater is limited all over the world (quoted from http://www.satnavi.jaxa.jp/project/gpm/tech/goal.html)

Chapter 5 Water: Problems and Conflicts   37

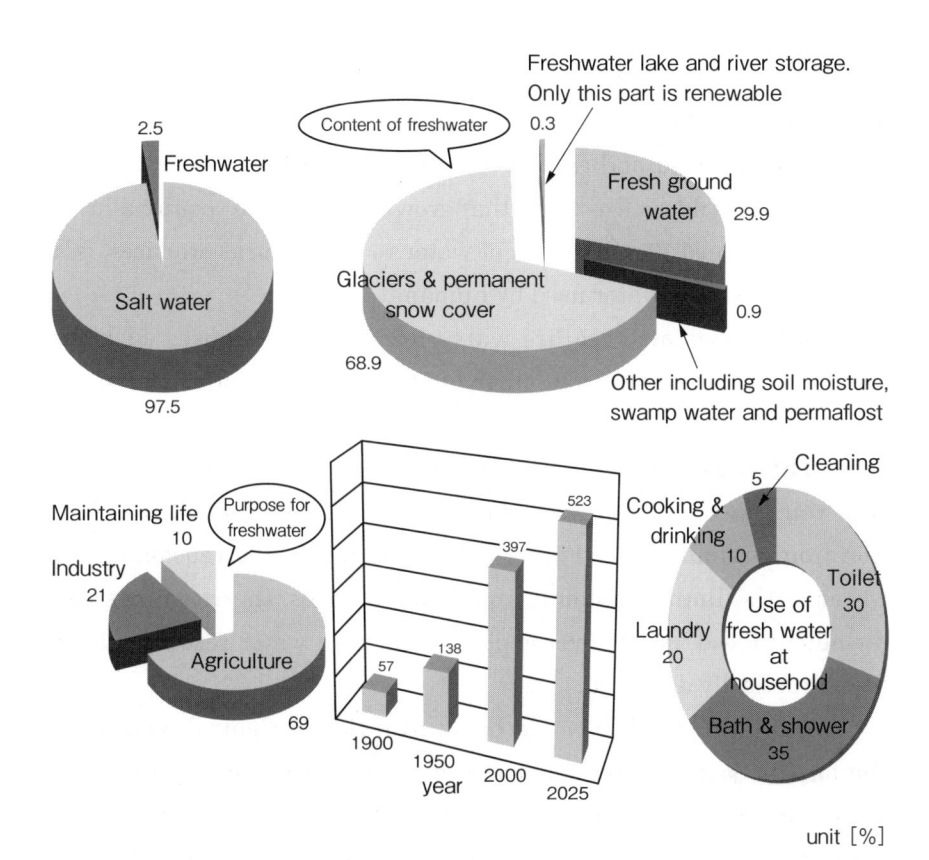

Figure 5.1 continuation

Only limited number of species such as halophytic plants and microorganisms can be tolerant to such conditions and majority of creatures on the earth need fresh water. Fresh water is only 2.5 % of the total amount of the water which exists in this planet. Seventy percent of the fresh water is not used and stay as glaciers and permanent ice and snow.

How do we use the water? Quenching thirsty, cooking food, washing cloths, bathing, flushing toilets, etc. Look at Figure 5.1 again. For our ordinary life, flushing toilets and bathing take a large proportion of water uses. A Japanese uses 313 litters of fresh water daily and on the virtual water which was discussed in Chapter 2, an additional 1,460 litters of water is consumed to support each of us. Japan imports food and energy, and also virtually we import a lot of water! We should realize again that growing crops also require a lot of water and indeed the allocation of water to agricultural practices, is a large proportion of water used by humans.

Industries also require water for various purposes such for cleaning, cooling, mixing and purifying for processing industrial materials.

The reusable water for humans come only from large water reservoirs such as fresh water lakes and rivers with small contribution from ground water (Figure 5.1). Importantly, the reusable fresh water is the limited amount and without cares, the resources can dwindle. Indeed, due to over-exploitation, fresh water resources have been drastically damaged.

A typical example would be on Aral Sea at Central Asia. With the former Soviet Union policy and implementation, irrigation-based agriculture practices have been promoted for food, feed and industrial materials such as cotton. Lake water such from Aral Sea was the primary source and canals have been constructed for the irrigation purpose, and consequently a lot of fresh water has been consumed without reusable ways (**Figure 5.2**).

Crop growing depleted a lot of soil fertility because of mismanagement and salinization of soil become prominent with the heavy dependency of ground water for the irrigation.

Salinization took place with the concentration of saline by

Note:
Aral Sea「アラル海」

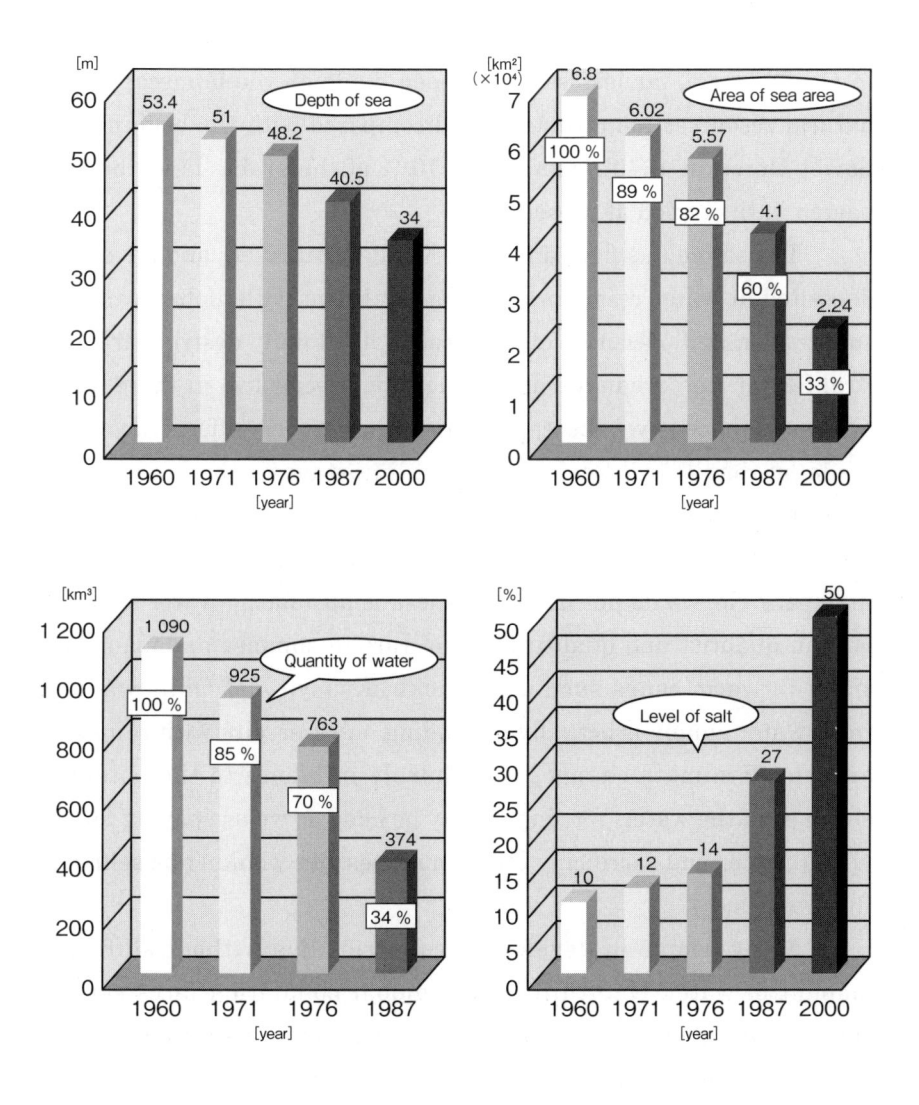

Figure 5.2 The most serious problem for water in Aral Sea
(modified from http://www.worldwatercouncil.org/ and http://www.eorc.nasda.go.jp/imgdata/topics/2004/tp04213.html)

capillary motion with sucking a lot of water up from deep wells. Furthermore, with the change of Soviet Union to the independent States in Central Asia, far less care has been made on the landscapes and natural resources management, and compared with the beginning of Soviet Union in 1940's, more than 70 % of the arable lands become barren in the region at present.

The saline has flowed into the fresh water rivers and have been deposited at water reservoirs such as Aral Sea. With global warming, salinization and over-uses of the fresh water totally destructed Large Aral Sea and no original domestic aquatic species live in there. However, small hope is yet there, protection and recovery efforts have been made on Small Aral Sea, and less saline stays in the lake and its water level have become higher.

Do we have enough water for our ordinary life and development in Japan? No, we do not have. Japanese is also facing water shortage both in quantity and quality. It is not rare in summer in Japan especially in warm zones such as Shikoku, Kyusyu and Okinawa. Our fresh water supply is actually depending on reservoirs with rain falls, and global climate changes can sharply influence to the potential shortage of the water. Yet, Japan may be okay as we are paying for the virtual water and sacrificing other countries, but wouldn't be selfish in that way?

Many people in developing countries live without sufficient amount of water. One third of the global population faces serious shortage of water. Even when they have access to water, it may be polluted or far apart from the quantity with which the folks ought to live with healthy conditions. Sanitation relates substantially to the quality of water, and babies and infants in developing countries die or live with very bad hygiene conditions due to difficulty to access clean water.

Chapter 5 Water: Problems and Conflicts

Water is the reason for domestic and international conflicts and problems. Indeed it has been the one throughout the human history. Stockholm Environmental Institute maps conflict and problematic sites and visit the websites for further information. Indeed, there are many locations in the world with inherent problems.

For example, Turkey made up a huge water reserve by regulating flow of Euphrates River, and Syria and Iraq face the serious water shortage, where already they have water supply problem. Mekong River flows from Tibet, Yunnan Province of China, Laos-Myanmar border, Cambodia then to Vietnam. Pollutions made at an upstream area, can concentrate and cause troubles at populated downstream regions such at Vietnam then to South China Sea, then to Pacific Ocean.

Water is the international concern. UN Millennium Development Goals set up eight major targets. These are not only for developing countries but also apply to the affluent societies such as Japan to really incorporate for the policy implementation and daily life of individual citizens.

The goals contain water related issues such as reduction of mortality of children, reduction of epidemic diseases and environmental sustainability. So that we should make sure that water is one of fundamental resources for human survival and coexistence with the nature.

Notes :
Euphrates River「ユーフラテス川」
Yunnan Province「雲南省」
UN Millennium Development Goals「国連ミレニアム開発目標」

Exercises

1. つぎの 1 ～ 52 の語に対応する英単語または英熟語を本文から選び出して書き込み，また発音しなさい（動詞は原形を記入しなさい）。

	Japanese	English		Japanese	English
1	紛争		19	再利用できる	
2	存在		20	貯蔵施設	
3	改善/向上		21	配分	
4	継続性		22	実行	
5	不可欠な		23	灌漑	
6	生き物		24	収奪する	
7	塩水		25	塩性化	
8	塩類		26	肥沃度	
9	耐塩性の		27	顕著な	
10	微生物		28	依存	
11	耐性		29	濃縮	
12	乾きを癒す		30	毛管現象	
13	流す		31	吸い上げる	
14	現実視する		32	井戸	
15	配分		33	景観	
16	実践		34	蓄積する	
17	割合		35	破戒する	
18	精製する		36	国内の/家庭の	

Chapter 5　Water: Problems and Conflicts　　43

	Japanese	English		Japanese	English
37	水生の		45	公衆衛生	
38	回復		46	衛生	
39	不足		47	取り込む	
40	降雨		48	削減	
41	汚染		49	致死率	
42	濃縮する		50	流行性の	
43	上流		51	根元の	
44	下流		52	共存	

2. つぎの各文が本文の内容と一致するものにはT(True)，一致しないものにはF(False)を，文末の（　）に記入しなさい。

（1） Japan does not have any water supply problem at all. (　)

（2） The reusable water is only 2.5 % of the water on this planet. (　)

（3） There is no global water supply concern as we have a lot of rain falls and as sea water can be converted. (　)

（4） Euphrates River region has water conflicts. (　)

（5） Japanese use water in other countries. (　)

3. つぎの日本語の各文を（　）の中の語を用いて英語の文にしなさい（必要があれば単語を適切な形に変換しなさい）。

（1） Stockholm Environmental Institute は世界中の水問題と紛争を編集し，地図にしている。(integrate, map, problematic)

（2）乳幼児の死亡率を下げるためには，きれいな水の十分な供給が必要である。(infant, mortality, reduction)

（3）土壌の塩性化は、地中深くの井戸から水をくみ上げることによって起こった毛管現象と塩類の濃縮に起因すると見なせる。(deem, capillary motion, attribute to, suck up)

（4）農業は大量の水を必要とし，都市生活での需要を圧迫している。(exert pressure on, practice, demand)

（5）UN Millennium Development Goals は八つの目標を掲げており，水は成功のための重要な要素である。(raise, component, attain)

4．つぎの各問いに英語で答えなさい。
（1）How can we share water resources globally?

(2) How can you make yourself contributed for water protection?

(3) Can destroyed water resources be recovered?

(4) Why did Central Asia region get the salinization and water shortage problems?

(5) Can we make use of sea water for various purposes?

5．つぎの日本語課題を英語で議論してください。
（ 1 ）日本は清浄で十分な量の水を国民一人一人が本当に享受できているでしょうか？　また，今後現状は維持できるでしょうか？

（ 2 ）トイレ，入浴，洗濯などで毎日多量の水を日本人は使っています。水の不足している世界各国と比べわれわれは水資源の重要性と節約を理解し，具体的な行動を示しているでしょうか？

Chapter 6   Waste and Domestic Garbage

An Average Japanese produce 1.1 kg of waste and garbage daily. Why do we produce that much? Where those rubbish stuffs go? How are they handled?

The garbage come from kitchen, is a big trouble in urban areas. Those are dumped or burnt at intensively designated sites but cause environmental and sanitary concerns. At rural sides, those stuffs could be dealt by secondary used as feeds to domestic animals or transformed as compost at ease.

However, nevertheless where people live in Japan, we know that there are a lot of fresh food becoming fossils at freezers, brand new edible stuffs left-over in refrigerators, and uneaten portion of food in each meal. Also we have over-supplies at stores such as unsold lunches at convenience stores. On the other hand, there are more than eight hundred million people in the world who are exposed always to hunger, and many of them are under serious starvation (**Figure 6.1**).

Affluent societies such in Japan should think about more sophistication rather than becoming satisfied with abusing precious food resources. Indeed, natural resources wise, Japan does not have affluence at all but scarce of many industrially and daily items such as food, metal and energy sources. We should avoid ruining the food, which is one of simple but effective actions to reduce garbage and wasting energy used in transportation, handling, processing and storage.

Promotions are made by local governments and civilian organizations as well as individual efforts of citizens: recycling of resources such as cans and pet bottles coming from dwindling natural resources, reuse of items by simple processing such as washing and disinfection

Chapter 6 Waste and Domestic Garbage **47**

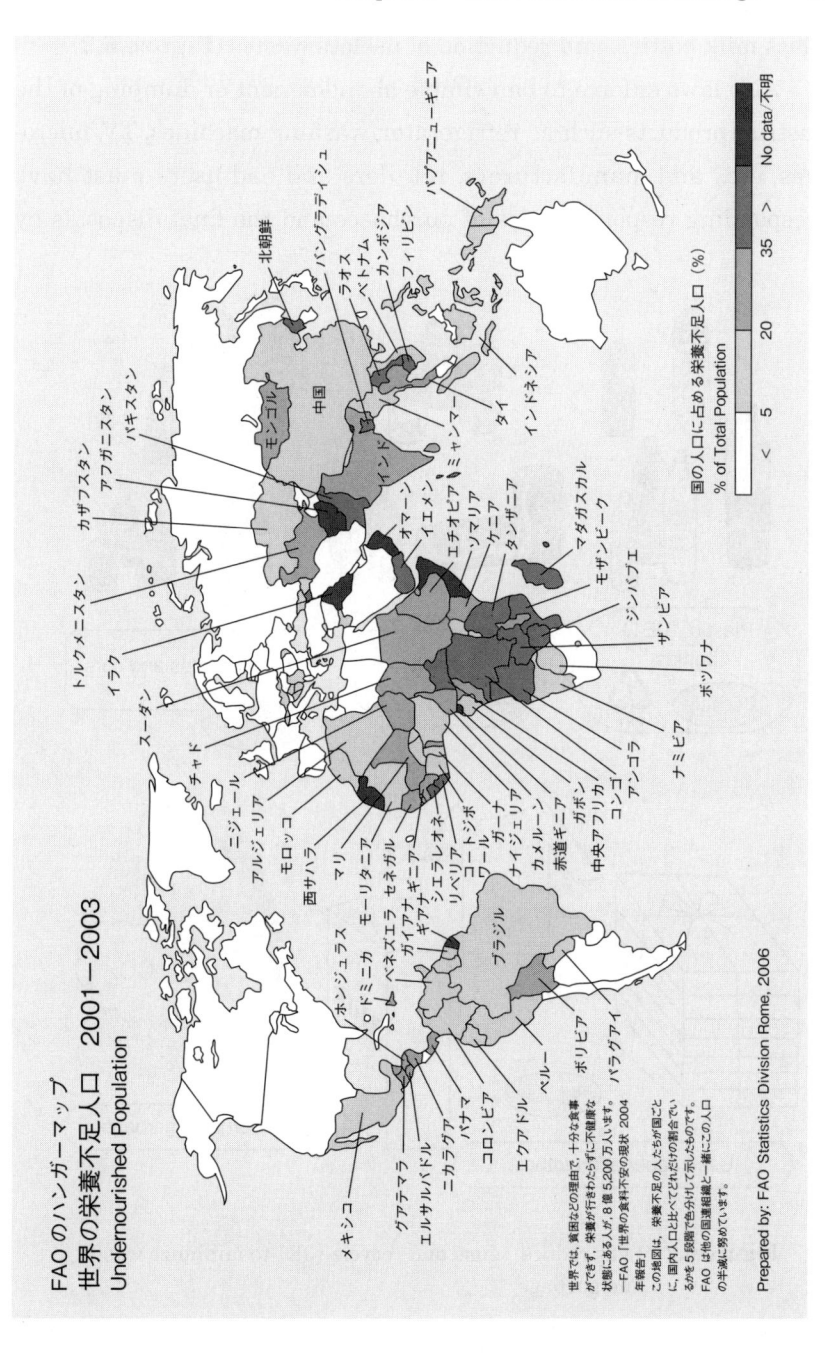

Figure 6.1 Still 852,000,000 people are hungry and starved

48 Unit B Natural Resources

of glass milk bottles, and reduction of useless wastes (**Figure 6.2**).

Also laws enforce to ban simple abandonment or dumping of the industrial products such as refrigerator, washing machines, TV, microwaves, etc., and manufacturers, retailers and end-users must have corresponding responsibilities of purchases and the final disposals by

Plastic (PET) bottles — Recycle

CANS — Recycle

Glass bottles and Jars — Recycle or reuse

Used paper and clothes — Recycle

Burnable garbage — CO_2

Figure 6.2 Let's reduce, reuse and recycle (3R) to minimize waste and garbage

Chapter 6 Waste and Domestic Garbage

fees and also in handling procedures. When automobiles are purchased, also disposal fees are included.

The disposal issues are not only on such a large items but also on small materials such as plastic bags furnished such by supermarkets and convenience stores. These plastic bags and wrapping materials compose sixty percent of the garbage coming from Japanese households. Unless it is burnt, there is no way to deal with the trash: we cannot bury or pile up somewhere which can cause environmental degradations.

Burning produces carbon dioxide and potentially more pollutants. Reusable "my bag" shall revive, which actually used to be at old time in Japan for daily shopping especially by housewives. Only with the effort, seven percent of your garbage can be reduced. Suppliers, retailers and consumers shall consider reducing packaging on daily purchased items and also on excess gift-wrapping.

Japanese buys up 40 % of the global shrimp products. Again more than half of the imports are spoiled or badly handled due to over-supply and ending to garbage. Many shrimps and prawns are cultured at the coastal areas of South and Southeast Asia. To meet with the market demands, the shrimp farms ruined a lot of mangrove vegetations which maintain coastal ecosystems.

The farms also polluted environments further by waste such from feed. Furthermore, to assure mass-production without diseases on the harvest, antibiotics overdosed often. Also lobster species are harvested from nature and species population is decreasing.

The processing facilities of the shrimps also produce enormous amount of stinky garbage from the raw materials such as heads and crusts. Demanding shrimps cause a lot of chain reactions to the environments at the production sites and global concerns as well as spoiled food stuff.

Natural resources are scarce in Japan, however, Recycling has also pitfalls. Aluminum cans can be recycled only at the rate of 30 to 40 %, first of all. For the purpose, a lot of energy should be spent for melting items and production of new items. Unrecycled parts become wastes and heat generated during the process for recycling also help the global warming.

A claim can be made, however: a lot of transport and processing industries would lose their business by reducing such consumable packing items and by decreasing long distance cargo services. People at developing countries also make income by selling the food stuffs to overseas.

Those are true, but to challenge the natural resources dwindling, global warming and real sustainability, a paradigm shift should be of consideration on the environments by each of people to be patient at some degree as far as survival of individuals guaranteed.

There must be a lot of social and ethical discussion to be made to alleviate the economic concerns, however, it is obvious that people living in bad environment cannot be wealthy and sustainable, thus environmental protection is the key also for the humanitarian issues at a long time frame.

How can we cope with all the rubbishes? It is valuable to care about the environments and be "a green consumer". Do not buy excess, purchase durable items even when it may cost higher, choose environmentally friendly products such as less wrapping and recyclable ones, bring own bag instead of the plastic ones, and stick to locally available food materials on season. Think global environments and act locally.

Note :
humanitarian「人道的な」

Chapter 6 Waste and Domestic Garbage

Exercises

1. つぎの 1 ～ 50 の語に対応する英単語または英熟語を本文から選び出して書き込み，また発音しなさい（動詞は原形を記入しなさい）。

	Japanese	English		Japanese	English
1	廃棄物		19	満足する	
2	生ゴミ		20	貴重な	
3	役に立たないもの		21	破壊する	
4	もの		22	輸送	
5	都会		23	取り扱う	
6	田舎		24	加工する	
7	放棄する		25	貯蔵	
8	方向付ける/指名する		26	振興	
9	衛生		27	激減する	
10	堆肥		28	再生（する）	
11	変換する		29	再利用（する）	
12	化石		30	削減する	
13	食用可能な		31	消毒	
14	露出する		32	強化する	
15	空腹		33	放棄	
16	飢餓		34	製造者	
17	裕福な		35	販売者	
18	洗練		36	末端利用者	

Unit B Natural Resources

	Japanese	English		Japanese	English
37	供給者		44	梱包	
38	購入する		45	包装	
39	供える		46	エビ	
40	ゴミ		47	海岸	
41	埋める		48	抗生物質	
42	積み上げる		49	欠点	
43	再登場する		50	溶かす	

2．つぎの各文が本文の内容と一致するものには T(True)，一致しないものには F(False) を，文末の（　）に記入しなさい。

（1）We cannot reduce garbage from our daily life. (　)
（2）Recycling cannot be made at the full efficiency such on alminum cans. (　)
（3）Environmental concerns and sustainability are more important than being affluent in many items at a short term. (　)
（4）It is always important on balancing the environmental issues and immediate humanitarian aids at many places where they have acute problems such as natural disasters. (　)
（5）Japanese dose not buy much shrimps from foreign countries. (　)

3．つぎの日本語の各文を（　）の中の語を用いて英語の文にしなさい（必要があれば単語を適切な形に変換しなさい）。

（1）ゴミを裏山に廃棄するのは違法なだけでなく，環境倫理や社会配慮に欠如した愚かな行為である。(dump, ridge, illegal, environmental ethics, inconsiderate, idiot)

（2）食品は計画的に購入し消費すべきである。(purchase, consume, plan)

（3）飢餓に苦しんでいる人たちが世界中には常にたくさんいる。(starve, worldwide)

（4）車や電化製品は，手続きなしに勝手に廃棄してはいけない。(abandon, prohibit, domestic electric items, procedure)

（5）家庭ゴミは，個人個人の配慮で軽減できる。(individual, consideration, decrease)

4．つぎの各問いに英語で答えなさい。
（1）Is it burden for the producers on Japanese demand on shrimp?

(2) What is your effort to reduce the garbage from your home?

(3) Should we participate in recycling although not all materials can be used?

(4) How can you make globally sustainable with such an enormous amount of wastes in the world?

(5) Should we buy brand new cars although global warming is serious, automobiles exhaust a lot of carbon dioxide and they accelerate the problem, and finally cars will be a huge rubbish once I abandon?

Chapter 7   Energy Demands and New Sources

Energy demands are skyrocketing globally with industrial development, transportation of bulk amount of materials and commodities, human movements locally and internationally and requirements for life improvement at local communities. Motor vehicle use is one of major reasons for energy demand, and simultaneously it is also one of the major causes on the global warming by the emission of CO_2. Everybody would like to have a convenient, healthy and safe life. Electricity is cardinal to provide such an occasion, especially as to keep life-lines for the urban life.

Yet many people use firewood for heating and cooking as primary energy sources, and over-exploitation of tree cuttings caused the forest damage and deforestation which is also the cause on environmental problems, and has association with global warming. It is yet the headache even when diverting such uses to other sources of energy, as we have limitation of energy supply for all demands in the planet.

We know that fossil-based energy sources such as petroleum will be consumed up in 30 to 50 years, upon how demands and actual uses are made in the coming future. Now there are screaming needs on finding alternatives globally.

Nuclear power is one to move on, however, management and infrastructure as well as human capacity are pitfalls to apply to many countries in the world. It is not an old memory that Japan had troubles at a major atomic-energy electricity generation facility at Niigata where there was a huge earthquake in 2007, and the atomic energy plant was damaged and some concern came up with the irradiation release.

IAEA is very skeptical on several countries where they may have abused uses or mismanagement toward catastrophes.

Then how do we manage the issue? Nuclear fusion would be dreamt but not yet actually achieved. Solar, wind, sea tide, ground-heat would be option but they could be used only locally and often do not apply for many occasions at many regions. Each area of the world should consider practical methods to generate own energy. Or less energy uses and go back to the Stone Age?

Plants can provides clean and reproducible energy (**Figure 7.1**).

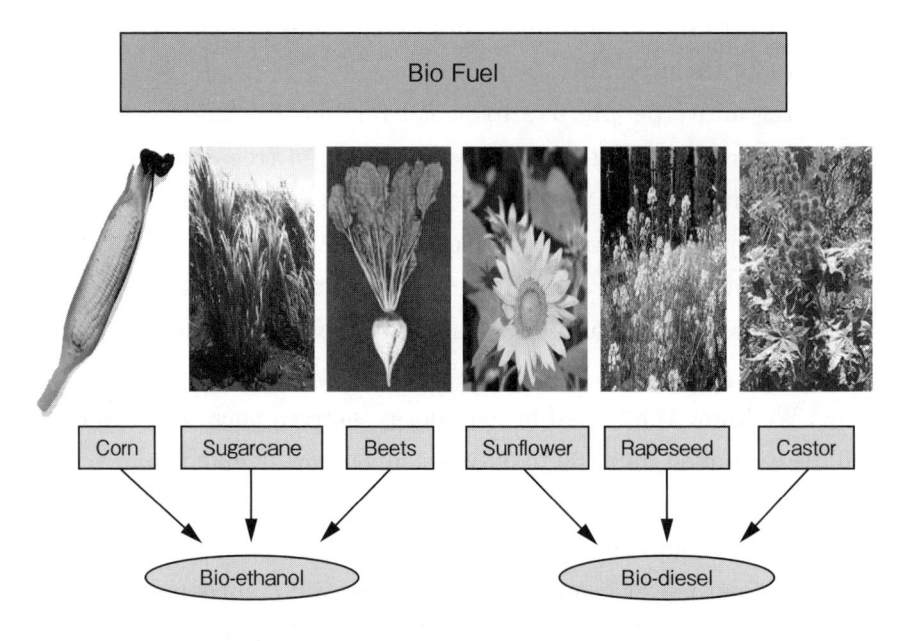

Bio Fuel

Corn, Sugarcane, Beets → Bio-ethanol

Sunflower, Rapeseed, Castor → Bio-diesel

Figure 7.1 Bio-fuel production has been accelerated

Note :

IAEA：International Atomic Energy Agency「国際原子力機関」

Chapter 7 Energy Demands and New Sources 57

During the global warming discussion at Kyoto Protocol and IPCC, those plant-based energy sources are regarded as low emission materials, and they are indeed promoted instead of combustible fossil materials. Bio-ethanol and bio-diesel are the converted sources.

Bio-ethanol can be made of sugar or starchy materials. Bio-diesel are processed from plant oils squeezed from botanical seeds rich in fatty acids or lipids etc. Imagine that cars can run with cooking oils!

Now the international prices of fossil energy sources are boosting even daily. And it is common understanding that sooner or later, the petroleum will be run out if humans continue using it. It is obvious that we need finding alternatives sources of energy. Ethanol can be provided from sugar-producing plants such as sugarcane, sugarbeet, maize and root and tuber crops like sweetpotato, cassava and potato. Brazil has already success of the large scale industrial production and utilization of bio-ethanol from sugarcanes since early 1970's, and the country is the leading nation on the technology and overall regulatory and social systems as well as common understanding of consumers.

Developed countries like Japan has not been careless in the past 40 years by forecasting the shortage of petroleum and the next generation of energy, however, many of developed countries had been indulged in the low price of petroleum up to recently, mainly because of economic reasons. But now paradigm has been shifted very rapidly.

The energy prices have been skyrocketing and the costs for

Notes:

Kyoto Protocol: Kyoto Protocol to The United Nations Framework Convention on Climate Change「京都議定書,http://unfccc.int/resource/convkp.html」

IPCC: Intergovernmental Panel on Climate Change「気候変動に関する政府間パネル,http://www.ipcc.ch/」

transportation and industrial processes which require energy, increase tremendously. Due to increasing interest and use of these crops for energy purposes, the prices of food and feed materials have been increasing for the past few years.

Poor people at developing countries now claim that they cannot buy food at a cheap price with the energy diversion trends of food crops: Mexicans worry about life without tortilla which is made of corn flour, and even Japanese delicacy sweet-stuff manufacturers are obliged to increase the retail prices of the cakes and snacks because of sugar price increases.

Food stuffs are supplied with higher prices, protests and riots have been taken places already in many places in the world as the consequent difficulty in living with the expensive commodities and energy supply limitation. It is not visible much, however, it will be, unless global collaboration is made to balancing the energy prices and food availability (**Figure 7.2**).

There are a lot of challenges to be made on science and technology associated with bio-energy. Secondary ethanol processing can be obtained by the help of chemical decomposition of cellulose which is the major compound to form up plants and also the basic materials for paper production. If this become practical, a lot of ethanol supply is guaranteed.

Further lignin which is the frame for plant architecture especially on trees, can be processed to make cellulose then to ethanol production. For these processes, yeasts and some bacterial species help the fermentation and genetic engineering is applied to enforce the capacity. Furthermore, industrial processing facilities are being considered for elaboration to maximize ethanol production and reduce the requirement of energy investment on the energy production.

Theoretically and experimentally, those processes are possible.

Chapter 7 Energy Demands and New Sources 59

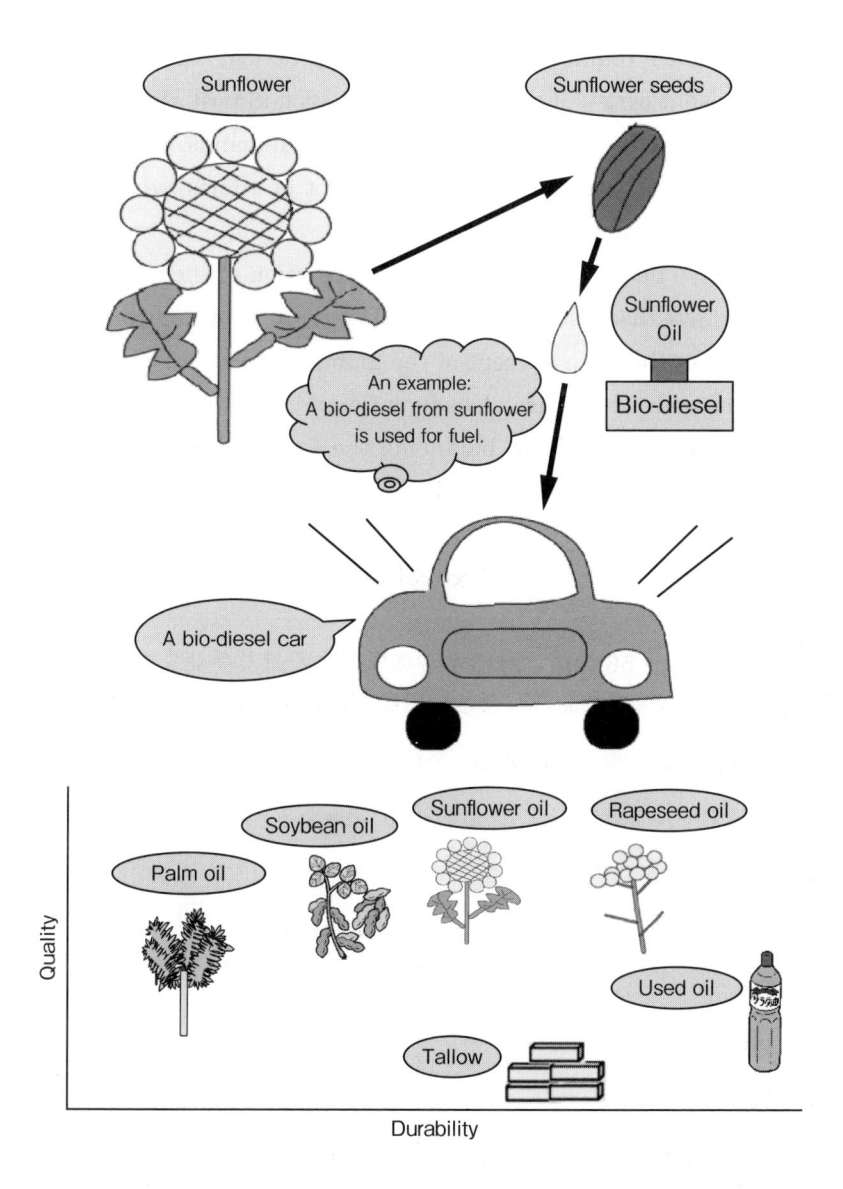

Figure 7.2 Many sources of bio-diesels can be used to alleviate the greenhouse effect.

However, a lot of basic research, application studies and industrial scale testing are required prior to efficient practical uses. Many scientists at present over-sell the potential and even almost are telling a lie that they can do it quickly available for large demands. Those technology development and application need a lot of efforts, which require rather a medium to long time frame.

We also have to be careful on the diversion of the crops to energy. As we have discussed in Chapter 6, especially on Figure 6.1 of the chapter, More than ten percent of the global population, which is over 800 million individuals are already suffered constantly from huger or even from starvation. Thus, balancing the food and energy sources is essential for the future sustainability issue.

Exercises

1. つぎの 1 ～ 56 の語に対応する英単語または英熟語を本文から選び出して書き込み，また発音しなさい（動詞は原形を記入しなさい）。

	Japanese	English		Japanese	English
1	飛躍する		10	過剰開発	
2	輸送		11	重要な	
3	物品/食料		12	森林破壊	
4	移動		13	転用する	
5	自動車		14	化石	
6	排気		15	消費する	
7	場合/出来事		16	わめく	
8	生命線		17	石油	
9	薪		18	原子力	

Chapter 7　Energy Demands and New Sources

	Japanese	English		Japanese	English
19	原子力		38	利用	
20	放射線		39	予測する	
21	放出する		40	気ままに楽しむ	
22	懐疑的である		41	大変な	
23	悪用する		42	転換	
24	管理不備		43	トルティーヤ	
25	大災害		44	粉	
26	核融合		45	優美な/お上品な	
27	達成する		46	小売り	
28	太陽の		47	暴動	
29	潮流		48	分解	
30	石器時代		49	セルロース	
31	再生できる		50	物質	
32	振興する		51	リグニン	
33	可燃の		52	構造	
34	増加する		53	発酵	
35	デンプン		54	遺伝子工学	
36	脂肪酸		55	改良	
37	脂質		56	投資	

2. つぎの各文が本文の内容と一致するものには T(True), 一致しないものには F(False) を, 文末の (　) に記入しなさい。

(1) Mexican folks do not have purchase of tortilla with the energy price increases. (　)

(2) We have no conflict interest ofon the use of crops for energy. (　)

(3) Brazil has already achieved the systematic production, processing and uses of the bio-ethanol. (　)

(4) Firewood is yet primary energy sources for many people. (　)

(5) Bio-diesel can be produced from any botanical seed. (　)

3. つぎの日本語の各文を (　) の中の語を用いて英語の文にしなさい (必要があれば単語を適切な形に変換しなさい)。

(1) 彼の研究の成果はテロリストに悪用された。(achievement, abuse, terrorists)

(2) 可燃エネルギーを使えなければ, われわれは石器時代の生活に戻ってしまう。(combustible, Stone Age)

(3) 原子力による発電は大量の電力を供給できるが, 管理は安全を担保するために重要な課題である。(Atomic energy, management, cardinal)

Chapter 7　Energy Demands and New Sources　**63**

（4）遺伝子工学により植物の代謝機能を改善し，脂肪酸を大量に産生させることは可能である。(metabolic functions, elaborate, fatty acids, facilitate)

（5）エネルギー資源の選択肢を開拓することは，人類の未来にとって重要な投資である。(exploit, options, investment)

4. つぎの各問いに英語で答えなさい。
（1）Is the bio-energy efficient for uses?

（2）How can you balance the use of corn for food, feed, industrial materials such as plastic and bio-energy?

（3）How do think engaging in research on bio-energy?

(4) We should ban atomic energy?

(5) Use of firewood should be stopped?

5．つぎの日本語課題を英語で議論してください。
　（1）新エネルギー資源として，バイオエネルギーが喧伝されています。樹木のリグニンやセルロースが糖化されるための大規模工業化は，まだほど遠いかもしれません。エネルギーを求めるだけではなく，日々どのような努力がエネルギーの利用削減になるでしょうか？

　（2）ゴミを燃やしても，結局は量が減るだけで，二酸化炭素や酸化物が残ります。これらを利用できる科学技術はできないでしょうか？

Chapter 8 Rare Metals and Conflicts

We have many occasions to hear about a word, rare metal, in these days. What are the rare metals? Rare metals are a group of elements consisting of 31 different entities. They are industrially important materials for high quality products such as cellular phones. Their property furnishes difficulty in extraction, purification and concentration from raw sources and/or scarce in existing amount in this planet.

The following parenthesis contain the atomic symbol and atomic number, respectively. Among the rare metals, seventeen entities are called rare earth such as scandium, (Sc, 21), yttrium (Y, 39) neodymium (Nd, 60) and terbium (Tb, 65). The latter two belong to Lanthanide series.

The Lanthanide series can be divided into two groups: cerium and yttrium. The element property of Lanthanide is similar each other, and because of the close similarity, it is complicated to conduct the isolation and purification of a specific element from the raw mixed materials from natural mines. The pure substance of these elements is easy to get oxidized.

In contrast to the rare metals, heavy metals such as copper (Cu, 29), antimony (Sb, 51) and zinc (Zn, 30), are called as base metals, and there are over ten times demands in quantity at global industries

---

Notes :
scandium「スカンジウム」
yttrium「イットリム」
neodymium「ネオジム」
terbium「テルビウム」
Lanthanide series「ランタノイドシリーズ」

compared with the rare metals. However, due to the skyrocketing growth in information technology related industry, the demands toward the rare metals will be also increasing tremendously.

Then, we shall know further about the rare metals. They have common properties such as heat tolerance, resistance to corruption, magnetic capacity and fluorescent activity. These materials also help smaller-sizing, reducing weight and higher performance of industrial products as well as energy-use efficiency. Japanese high technology industry had success with the rare metals, and there is no further growth without them.

Compounds and alloys made from them are used for ignition alloys, lenses, fluorescent items, laser sources, permanent magnet and base materials for computer semiconductor. Ceramics containing the rare earth can have high temperature superconductive capacity and are employed at different advanced industrial products. Then, you can imagine how they are important for your life as they are used for almost everything around you; televisions, cellular phones, computers, eyeglasses, room illuminating fluorescent lights etc.(**Figure 8.1**).

Platinum (Pt, 78) is also a rare metal and it is used for essential catalyst for gas-phase remediation. Automobile industry consume the sixty percent of world demands on platinum. Each car needs three gram of platinum. The up-coming energy cell operated car requires about 80 gram per vehicle. How much does it cost at present on the price?　400,000 yen per car!

Gallium (Ga, 31) is used for LED in cellular phones. Since the

Notes：
skyrocketing「飛躍的に増加する」
gallium「ガリウム」
LED：light emitting diode「発光ダイオード」

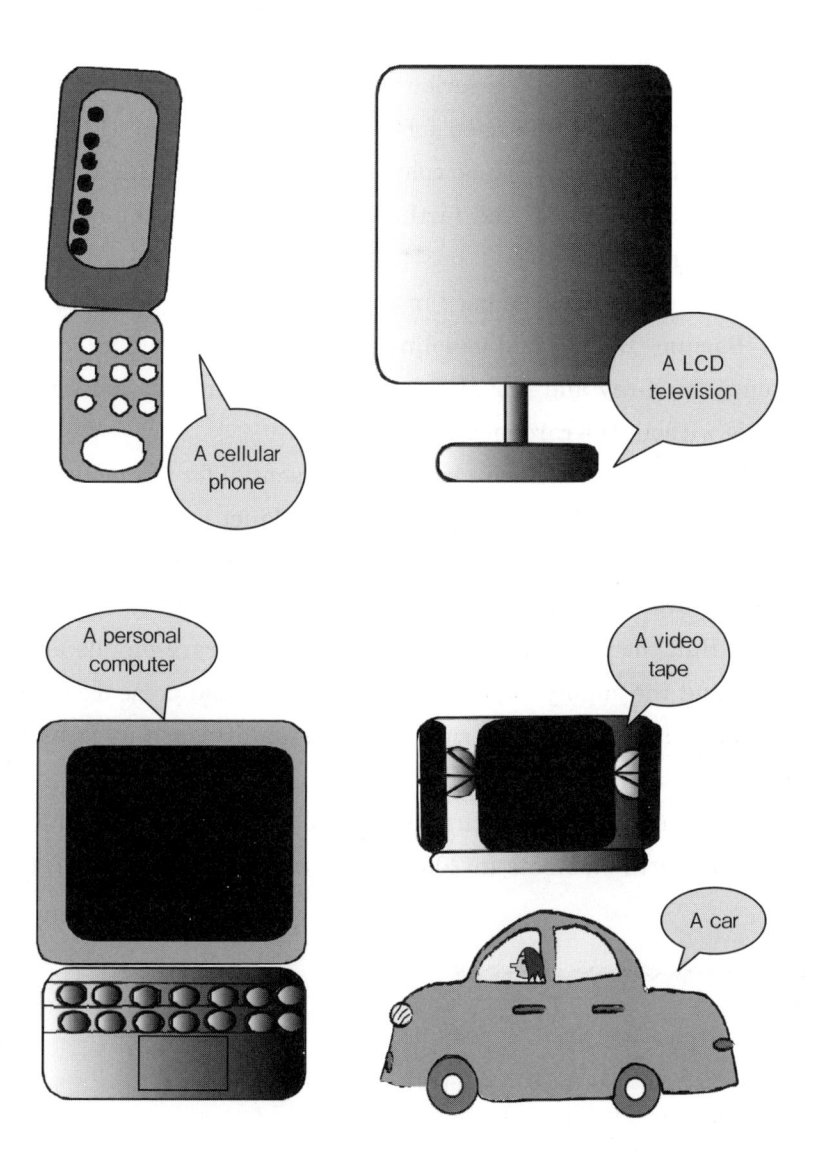

Figure 8.1 Examples for using rare metals

uses of the handy phones are very popular among many nations, and there will be more demands, especially populated countries such as India and China. There is a serious concern on the shortage of supply of Ga. Indium (In, 49) is used for liquid crystal display (LCD) and plasma TV. The element is also contained extensively in the materials such as solder and fuses, and furthermore, it is used also for alloys in dental materials.

Rare metals are also used in precious medical devices and practices. Barium (Ba, 56) and gadolinium (Gd, 64) are used as contrast reagents, for X-ray and for MRI, respectively. Then, can you imagine your life without the rare metals?

Now we know that rare metals are essential for the modern life. Base metals are widely distributed in this planet and the amount is relatively abundant to meet with the global demands. Also political and social systems support the international availability of the substances. But, the geographical distribution of the rare metals are restricted and the supply amount is limited. Often the presence of such rare metals are at the countries under instability such as local conflicts and wars.

For example, China occupied the 93.3 % of the global production of the rare earth elements in 2005. China also produces a lot of met-

Notes:
indium「インジウム」
liquid crystal display (LCD)「液晶画面」
plasma TV「プラズマテレビ」
barium「バリウム」
gadolinium「ガドリニウム」
X-ray「エックス線」
MRI：magnetic resonance imaging「磁気共鳴画像」

als as raw industrial materials: Eighty-five percent of global antimony production, fifty-four percent on indium, fifty-one percent on barium, respectively.

The global concerns exist on the metals from China: instability of the supply and price fluctuation. First let's look at base metals. In 2004 to 2005, the price of manganese (Mn, 25) became five times more. The domestic demands in China influenced the market value of nickel (Ni, 28), aluminum (Al, 13) and tungsten (W, 74) with 15 % and 10 % in rare earths on tax increase, respectively. China takes strategy to make uses of the rare metals to manufacture high-technology products as a key export entity.

Japan imports majority of its demand on the rare metals from China, and it is deleterious to depend on one country on supply in terms of resources security aspect as Japan has to compete for the resources and also products for export with China.

Depending on the elements, there are options on the resources suppliers. South Africa provides 78 % of the global production of platinum, 42 % of vanadium (V, 23), 43 % of chromium (Cr, 24) and 23 % of manganese, respectively. Russia also supplies 21 % of vanadium, 12 % of platinum, respectively. Molybdenum (Mo, 42) is rich in USA (34 %) and Chile (27 %). Chromium also distributed in India (19 %) and Kazafstan (19 %).

However, there are rigid regulatory systems on the exploitation of the metal resources by foreign investment and it takes more cumbersome in diplomatic negotiation and civilian business talks.

Notes :
vanadium 「バナジウム」
chromium 「クロム」
molybdenum 「モリブデン」

Furthermore, many nations with rare metals often face the domestic conflicts and regional wars. Some of African conflicts are mixed up with the natural resources availability, and the rare metals are often targets of the armed bloody fighting. These make more difficulty in short term access to the materials and negative direction for future long term development with the uses of the precious natural resources.

Since the access is limited and prices are high in rare metals, it is now important to reuse, recycle and/or substitute the rare metals like efforts have been already made on base metals such as aluminum cans. However, endeavors are needed for the recycling of the rare metals as high technology with cost reduction is essential for the purpose. It is costly and tedious, but a country like Japan does not have natural resources on rare metals, so that efforts and practical technology are encouraged for the recycling.

Furthermore, world will require more rare metals for various uses for development, and Japan shall have own strategy to conserve acquired rare metals and to further access to the raw materials in the supplying countries.

Exercises

1. つぎの 1 ～ 45 の語に対応する英単語または英熟語を本文から選び出して書き込み，また発音しなさい（動詞は原形を記入しなさい）。

	Japanese	English		Japanese	English
1	希少金属		4	製品	
2	元素		5	携帯電話	
3	実態		6	特性	

Chapter 8 Rare Metals and Conflicts

	Japanese	English		Japanese	English
7	備える		27	合金	
8	抽出		28	発火	
9	精製		29	半導体	
10	濃縮		30	セラミック	
11	乏しい		31	超伝導の	
12	括弧		32	照明する	
13	元素記号		33	プラチナ	
14	元素番号		34	地理的な	
15	希土類		35	不安定	
16	鉱山		36	変動	
17	酸化する		37	ニッケル	
18	銅		38	アルミニウム	
19	鉛		39	タングステン	
20	亜鉛		40	製造する	
21	多大な		41	危険な	
22	耐熱性		42	資源保障	
23	耐腐食性		43	外交交渉	
24	磁石の		44	手間のかかる	
25	蛍光の		45	獲得する	
26	化合物				

2. つぎの各文が本文の内容と一致するものにはT(True)，一致しないものにはF(False)を，文末の（　）に記入しなさい。

（1）China occupied the 93.3 % of the global production of the rare earth elements in 2005. (　　)

（2）Depending on the elements, there are no option on the resources suppliers. (　　)

（3）Many nations with rare metals often face the domestic conflicts and regional wars. (　　)

（4）It is now important to reuse, recycle and/or substitute the rare metals. (　　)

（5）Recycling of the rare metals is costly and tedious. (　　)

3. つぎの日本語の各文を（　）の中の語を用いて英語の文にしなさい（必要があれば単語を適切な形に変換しなさい）。

（1）携帯電話の利用は世界的に増加しており，携帯電話の原料である希少金属の供給は世界的な懸案事項である。(sky-rocket, cellular phone, concern, supply)

（2）希土類は特性が似ている。(rare earth, element, similar, property)

（3）希少金属は，高次に加工された合金や医療用の機器にも使われている。(alloy, process, employ, medical devices)

（4）バリウムやガドリニウムは，X 線や MRI の投影剤に使われている。(barium, gadolinium, contrast reagents)

（5）資源開発分野への外資参入を規制する動きも活発化しており，その他の国との取引についても外交的な手段が試されるところである。(regulatory system, exploitation, diplomatic negotiation, complicated)

4. つぎの各問いに英語で答えなさい。
（1）What are the rare metals?

（2）What are rare earth elements?

（3）How are the rare metals used?

(4) What is the property of the rare metals?

(5) Does an automobile need rare metal?

5. つぎの日本語課題を英語で議論してください。
　（１）日本は，都市鉱山をもっていると認知しながら，リサイクルや再使用はあまり進んでいません。コストが制限要素である一方，社会のシステムの新規構築が必要ではないでしょうか？

　（２）日本は，世界各国の鉱物資源を本当に平和的に入手しているでしょうか？　間接的に紛争に加担するような支援をしていないでしょうか？　また，日本の消費者それぞれの需要がこのような紛争に遠隔で繋がっていないでしょうか？

　（３）ダイヤモンドやルビーなどが，軍隊の資金源として使われていることが往々にしてあります。加工流通業者は，このような由来の宝石を取り扱わないようにしていますが，消費者もこのような意識をもって不買等を考える必要がありますが，どのように bloody jewelry を避けることができるのでしょうか？

Unit C Environment

Chapter 9 Food Chain and Biodiversity

Poo is a small and mixed breed dog with white and black piebald fur. He is young, curious but a bit silly, which his family humans like. He is the introducer of the scenes in his backyard and precincts. Poo had lived at an urban side and now with his owner family, he lives at a rural area where is rich in biological diversity. He is very satisfied with his new territory with a lot of curiosity in exploiting for his neighborhood.

Poo's family members like various flowers blooming in the spring. Many tree and herbaceous plant species set colorful and smelling flowers. Honey bees, bumble-bees, various butterflies and some sort of flies are gathering around the flowers for nectar.

Poo is fond of observing that butterflies and moths lay eggs on the green plants. Poo also peeps that the eggs hatch and caterpillar worms feed on the greens, and predators on them such as killer bees, reptiles such as lizards and birds come around the flowers to catch the preys.

Poo's garden has plenty of trees with various taxonomic entities: needle-leaved and broad-leaved evergreens and deciduous broad-leaved ones with beautiful autumn tints. There are also many edible fruit trees, bushes and grapevines. He jumps into the shrubs when a lot of berries sets. He likes raspberries and blueberries as well as wild strawberries.

Birds eat majority of the plums and cherries in spring and persimmons and grapes in fall, which Poo cannot reach to the height of branches at which fruit ripen. The birds sometimes tease Poo by leaving excrements on the head of him.

With the living nature as a dog, Poo likes digging grounds. Also for his curiosity on the animal smells coming from small holes and soil mounts, he digs often around them. He founds eelworms and the tunnels made by moles. Each mole has own underground territory for keeping own food stuff such as eelworms and small soil-borne insects.

Poo is also fond of the small pond and marsh surrounding it in the backyard of his house. In marsh when water is plenty in spring, Japanese rice fish, medaka, make a fleet with its cohorts.

He also likes the segregation of the filial progeny from the red and ordinary dark color medaka parents: it follows Mendelian genetics. Medaka also have blue and white color bodies which are caused by the genetic property.

Poo knows that medaka has been an appropriate study material for inheritance but now it is endangered in Japan due to the loss of their habitats with the disappearance of old paddy-based agriculture.

Traditional paddy production system was an ecosystem: rice, Japanese barn millet as weed, arrowhead with its weedy relative, omodaka (*Saggitaria trifolia*), mud snail, loach, as well as medaka. Soil management with use of a diversity of rice varieties, has been easy to avoid inherent, identical and serious soil-borne diseases with the change of water flooding period at the growing time of rice and dried field condition while crops were harvested.

The system also provided a lot of food materials to wild birds such as Japanese crested ibis, herons, oriental stork and ducks. Also

Notes :
omodaka (*Saggitaria trifolia*) 「オモダカ」
Japanese crested ibis 「トキ」
oriental stork 「コウノトリ」

local people used those species from paddy for consumption.

It has been drastic changes: conversion of the paddy field to other uses or application of agricultural chemicals and change in the water supply system to paddy changed ecosystem and consequently many of the species are being endangered.

In the pond along with marsh, there are many aquatic plants. Poo enjoys water lilies with beautiful flowers in summer; lotus, water chestnut (*Trapa bispinosa*)and mizuaoi (*Monochoria korsakowii*), which are edible; and common reed (*Phragmites communis*) which is good for crafts, for maintaining landscape against erosion and also for phyto-remediation.

At such a water-front environment, there are also many living organisms making interactions. Dragon flies and its nymphs, water spiders, amphibians such as salamanders and frogs, snails, crawfish and a variety of fish including medaka. Again birds are primary predators to the small creatures. Also those aquatic living organisms make food chain: for example, small fish feed on planktons and algae, yago, the dragonflies' nymph can be a predator to small fish and aquatic insects, and the adult dragon flies can be prey for the birds.

Population size of a species can be maintained at a time frame, once when the food is abundant, species proliferate unless there are serious external influence such as diseases and when the food is dwindling then the species become scarce or rare. Poo feels that he would also be in the larger food chain if he lives in wild, while he is accom-

Notes :

water chestnut (*Trapa bispinosa*)「ヒシ」
mizuaoi (*Monochoria korsakowii*)「水葱（ミズナギ）」
common reed (*Phragmites communis*)「ヨシ」

modated by human as a domesticated companion animal.

Poo had seen a carcass of a cicada at a foot of a pine tree. In the next day, he has seen ants gathering around the dead insect, and the following day, he saw nothing there. The dead body is not a waste, it is an important sources for a life. Food chain and recurrent cycles of life forms are essential in perpetuating biodiversity. Beside that human influences should be definitely important to keep the healthy balance of the ecosystem.

He felt all living creatures connected each other and nothing is useless. A life depends on another one, like Poo and his human family members, and reincarnation may be true for all living creatures. Poo also contemplate that respect to any of life has the universal value beyond ethical and religious issues, while he fell asleep at a clover covered ground with warm spring sun light.

Exercises

1. つぎの 1 〜 63 の語に対応する英単語または英熟語を本文から選び出して書き込み，また発音しなさい（動詞は原形を記入しなさい）。

	Japanese	English		Japanese	English
1	食物連鎖		8	田舎の	
2	生物多様性		9	近郊	
3	雑種		10	縄張り/領域	
4	まだらの		11	調査する	
5	毛皮		12	蜜	
6	まぬけな		13	食肉種	
7	都会の		14	は虫類	

Chapter 9 Food Chain and Biodiversity

	Japanese	English		Japanese	English
15	トカゲ		36	タニシ	
16	分類の		37	ドジョウ	
17	常緑（の）		38	鷺	
18	落葉の		39	急激な	
19	針葉の		40	湿地	
20	広葉の		41	景観	
21	茂み		42	浸食	
22	低木		43	生き物	
23	紅葉		44	水生の	
24	枝		45	両生類	
25	排泄物		46	ザリガニ	
26	ミミズ		47	サンショウウオ	
27	モグラ		48	トンボ	
28	群れ		49	幼虫	
29	遺伝学		50	集団	
30	遺伝		51	豊富な	
31	絶滅危惧の		52	増殖する	
32	田んぼ		53	激減する	
33	生態系		54	少ない	
34	クワイ		55	稀少な	
35	ヒエ		56	家畜化された	

80   Unit C   Environment

	Japanese	English		Japanese	English
57	セミ		61	輪廻転生	
58	ゴミ/廃棄物		62	倫理的な	
59	永続させる		63	宗教の	
60	循環の				

2．つぎの各文が本文の内容と一致するものにはT(True)，一致しないものにはF(False)を，文末の（　）に記入しなさい．
　（1）Arrowhead is edible. (　)
　（2）Medaka is an endangered species. (　)
　（3）Lizards are amphibians. (　)
　（4）Crawhfish is a member of fish. (　)
　（5）Yago is the nymph of the cicada. (　)

3．つぎの日本語の各文を（　）の中の語を用いて英語の文にしなさい（必要があれば単語を適切な形に変換しなさい）．
　（1）プーは常緑樹の枝にいるトカゲを観察した．(observe, lizard, evergreen)

　（2）ニュージーランドなどの南半球の国では，紅葉は3月から5月ごろに見られる．(Southern hemisphere, see)

（3）メダカの体色は，メンデル式の遺伝をする。(Mendelian, inheritance, follow)

（4）生物多様性は地球温暖化と環境破壊で激減している。(dwindle, global warming, environmental destruction)

（5）食物連鎖の頂点に立つ生物は，その種を永続できる。(peak, perpetuate)

4. つぎの各問いに英語で答えなさい。
（1）What is difference between evergreen and deciduous trees?

（2）Are salamanders and lizards in the same taxonomic entity as they look alike?

(3) Can biodiversity be perpetuated without human efforts?

(4) Do you believe the reincarnation?

(5) Is ethics an important component in biodiversity conservation?

5．つぎの日本語課題を英語で議論してください。
（1）食物連鎖を人間に当てはめると，人類はいかにたくさんの生き物の存在や利用によって生存できているかが認知できます。生態系と食料生産の関わりを幅広く考えてみましょう。

（2）絶滅危惧種について，人類が人工的に生存を支援することは，意義があるのでしょうか？　投資する経済的な負担や先行きの不透明性などを考えると滅びてもいいのでは？　あるいは，これまでの急速な生物多様性の減少は人類の生産活動によるものであり，なにをもっても人類と地球が共存できるように，生物多様性の保全と持続的利用は，人類の優先対処事項でしょうか？

Chapter 10　Invasive Species and Crisis in Biodiversity

Natural migration of species have been taking places in the history of our planet earth and it is one of major speciation and evolutionary processes. Birds move from an island to other places, and even small butterflies can make a long distance flight over seas. Hard seed and vegetative propagules could float over oceans, and small seeds are carried by birds. These events were natural occurrences and with the long time frame of the life of the earth, they are healthy phenomena that did not make a drastic disturbance on the ecosystem and biodiversity.

But there were five major extinction spasms in the planet with natural catastrophes such as cataclysm, global freezing and other speculated reasons like a huge meteorite struck the earth which destructed environments. Biodiversity had been lost once upon a time then natural recovery had been made gradually by the change of environments with new species evolving from the climate and ecosystem changes.

Many scientists are now saying that human activities cause the sixth extinction spasm.
Those causes are:
1) Human destruction of ecosystems,
2) Overexploitation of species and natural resources,
3) Human overpopulation,
4) The spread of agriculture, and
5) Pollution.

Unfortunately, those are often irreversible and are unavoidable as far as humans live on the earth, but alleviation efforts shall be enforced to reduce the problems.

After the exploitation of the new continents such as Americas and Australia, and also with the rapid transportation development in the past two centuries, enormous amount of species had been mobilized intentionally and unintentionally by humans. There have been changes of the ecosystems by the introduction of exotic species, often they were drastic or destructive.

On the other hand, we shall also recognize that intentional movement of some species have positive effects for human life, such as exchange of many crop species between Americas and old continents.

Look at your ordinary tables: maize, tomato, potato, French bean (*Phaseolus vulgaris*), chili pepper and squash, originated in Americas. Those crops are essential materials for many culinary cultures at any places.

Can Italians imagine their dishes without tomatoes? How could Southeast Asians like Thais live without hot chili peppers? Soybeans are grown at enormous amount in USA and Brazil, which came from East Asia and now feeding and supporting Japanese demands. Canola, a modified rapeseed, is globally important and healthy edible oil sources, which is produced intensively in Canada, and again Japanese consumers depend on the import from there.

These Brassica species came from the old continent to Americas. Bananas and plantains (*Musa* spp.) can be seen any tropical and warm zones in the world, and often it is important staple food such as at Africa. *Musa* species have a large diversity and origin zone in Pacific such at Papua New Guinea to Southeast Asia.

Biodiversity become more uniform and the diversity per se is dwindling over the planet Earth compared with the past two centu-

---

Note :

Brassica「キャベツ，ハクサイ，カブ，ナタネ等の仲間」

ries. Active movements of humans from a continent to another continent and global warming are the cause to the drastic loss of biodiversity. Such an event is not rare also in Japan.

Old natural and farm-village sceneries which can be evoked in Japanese nursery rhymes, are not real now and lost in Japan. Medaka is endangered, domestic dandelions are extinguishing due to hybridization with exotic dandelions, Japanese monkeys also mix with Taiwanese ones which escaped from zoos. As discussed in the previous chapter, due to the change of even farm-land uses without destroying the natural environment, many of species such as oriental storks and Japanese crested ibis are endangered and lost, respectively. It is happening around your environment and unfortunately it is irreversible and no way to recover many of landscapes and species living there.

A few number of species become often abundant in our precincts. Many of them are exotic, did not derive from our homeland. Many of them had been intentionally introduced with specific reasons why did not come with ecological concern as the consequence of the deliberate release to the environment.

Examples would be blackbasses and bluegills in freshwater lakes and ponds (**Figure 10.1**), which were introduced for sport fishing. Catfish was introduced for replacing carp production in lakes such as Kasumigaura, once carp herpes virus infested the fish population and destroyed commercial production at many places in Japan. Those species introduced from abroad, become predators to endemic species such as varieties of bitterlings and freshwater shrimps, and

Notes:
black basses and bluegills 「オオクチバスやブルーギル」
carp herpes virus 「コイヘルペスウイルス」
bitterling 「タナゴ」

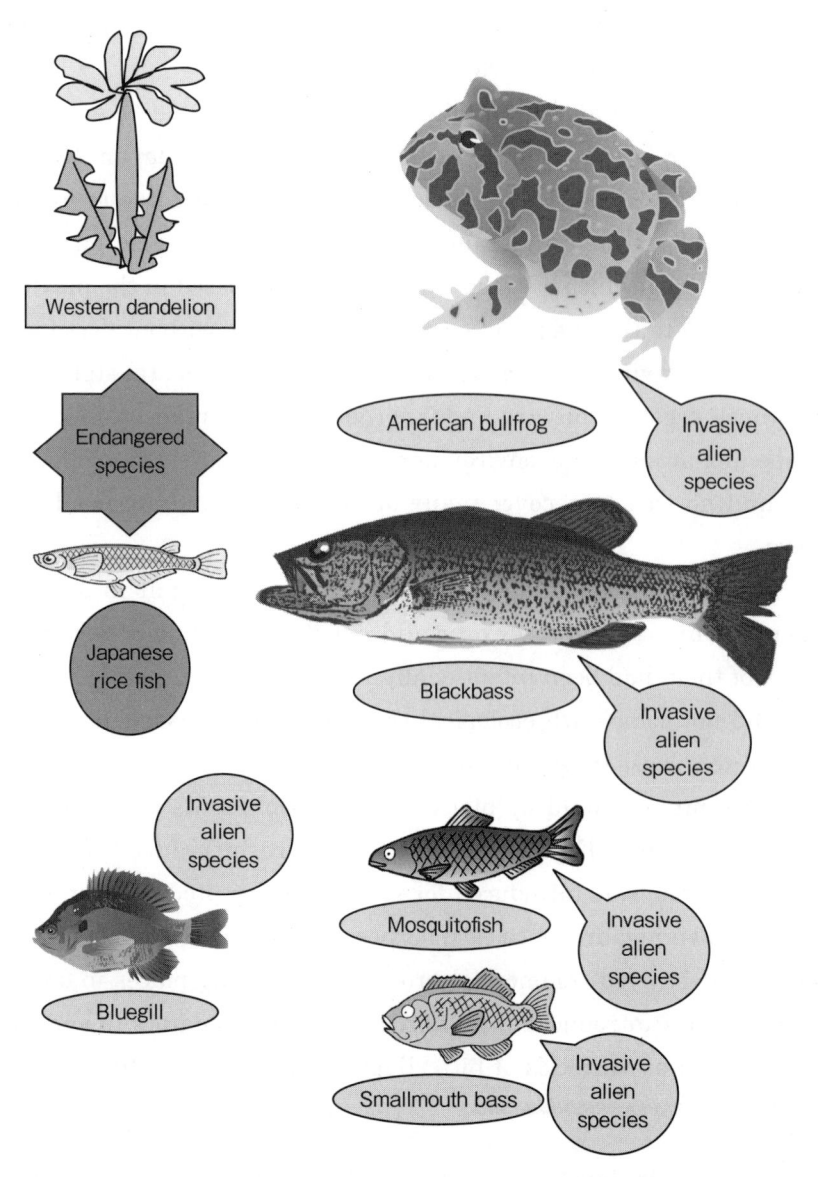

Figure 10.1 Examples of invasive and endangered species

Chapter 10 Invasive Species and Crisis in Biodiversity

those foreign species over-compete, damaged and even ruined the domestic species populations.

Due to the invasiveness of those species, Japanese traditional freshwater ecosystem had been changed drastically and often destroyed, by making many of endemic species become, extinct, endangered or threatened.

Lets look at vegetations around us. We often see Canada golden-rod (*Solidago altissima*). It is an herbaceous perennial plant. It was introduced from North America primary aimed for honey sources. However, it spread over Japan, and even the species competed with indigenous vegetation and destroyed many species with its allelopathic chemical substance, called cis-DME. It has high fecundity and pollen is insect-borne. Besides the sexual reproduction by seeds, vegetative propagation makes strong presence in the local vegetation. This is an invasive weed, and it is also problems in other regions such as China and Europe.

Water lily (*Eichhornia crassipes*) have beautiful flowers. It originated from South America. It can take over the aquatic ecosystem easily by destroying other domestic aquatic plant species by rapid proliferation and allelopathic substances. It is a globally problematic species listed in The World Conservation Union (IUCN).

Physella acuta, commonly called as European physa, or sakamakigai in Japanese, is a species of small, left-handed or sinistral, air-breathing freshwater snail. The small snail is another example. It came into Japan probably with algae and seaweed for hobby aquarium business. Origin may be regarded at Europe. Now it proliferates explo-

Notes :
Canada golden-rod 「セイタカアワダチソウ」
The World Conservation Union (IUCN) 「国際自然保護連合」

sively at elsewhere in Japan. Also it has very flexible adaptability to environmental degradations, so that it can thrive even at dirty sewage water conditions, and with its strong viability and fitness, it is used as an index species on the environmental pollution.

If the species cannot survive in the environment, it is really a creature-dead condition. Also it can be a vector of infectious or parasitic protozoa to human.

Many countries now have legal systems to avoid the environmental damages and destructions including Japan. Those laws regulate deliberate introduction of exotic species into the country even at a confined condition. There are two major aspects: 1) invasiveness in the environments and 2) pathogenic or parasitic to human and agricultural species.

The former case is often called invasive species law and the latter case is associated with quarantines with human, plants and animals. On the invasiveness concern, Japanese Environment Ministry makes campaign as: "No introduction, No abandonment, No spread of exotic species."

Then, let's think about our daily situation. Would it be okay to bring alien plant seed of very unique and beautiful flowers from foreign country when we go for sightseeing overseas? No. It may have invasiveness in Japanese territory, it may carry plant pathogen which could have devastating damage to crops, and what is more, international laws also regulate the country ownership of biodiversity. So that without formal processes on the species introduction, there are a lot of legal and practical problems.

Another exercise would be at your home. You like keeping aquariums at home with a lot of tropical fish such as catfish and guppies. Can it be release at a pond near by your house? No, unless scientific and legal statements are provided, it could cause full of social and ecological problems to your neighboring environment. Indeed, tropical fish like guppies are problems in the natural environment in Okinawa islands. Even as stated at the early section of this chapter, many of catfish from abroad can be invasive.

At last, a prejudice should be avoided: not all foreign species are invasive, however. And scientific information is not only the factor for the decision-making on the introduction of a exotic species. Precaution is always essential on the introduction of alien species into a new environment, and a logical and scientific risk assessments should be made on the possibility and degree of the invasiveness on a species.

Legal and regulatory systems must make a final judgment together with the biological and social scientific information on the matter. It is not only the matter of good or bad to environment, but also it could be balance of risk-benefit.

Then, an end-point decision can be made on the introduction of a foreign species into the country and on any deliberate release to an exotic environment.

Exercises

1. つぎの 1 〜 70 の語に対応する英単語または英熟語を本文から選び出して書き込み，また発音しなさい（動詞は原形を記入しなさい）。

	Japanese	English		Japanese	English
1	侵略性の		5	進化の	
2	危機		6	繁殖源	
3	移動		7	発生	
4	種の分化		8	現象	

Unit C Environment

	Japanese	English		Japanese	English
9	妨害/撹乱		31	風景	
10	消滅		32	喚起する	
11	症摩		33	童謡	
12	大災害		34	タンポポ	
13	地殻変動		35	交雑	
14	類推する		36	絶滅危惧の	
15	隕石		37	絶やす	
16	衝突する		38	絶滅した	
17	進化する		39	恐れがある	
18	気候		40	結果	
19	過剰な開発		41	懸案	
20	汚染		42	意図的な	
21	不可逆な		43	ナマズ	
22	軽減		44	壊滅させる/廃墟	
23	強化する		45	草本性	
24	膨大な		46	物質	
25	移動する		47	アレロパシー性の	
26	意図的に		48	生産力/多産	
27	調理の		49	花粉	
28	食用バナナ		50	虫媒	
29	減少する		51	生殖	
30	急激に		52	繁殖	

Chapter 10 Invasive Species and Crisis in Biodiversity

	Japanese	English		Japanese	English
53	植生		62	検疫	
54	水槽		63	病原（性の）	
55	爆発的に		64	廃棄	
56	適応力		65	規制する	
57	劣化		66	偏見	
58	生存する		67	在来の	
59	原生動物		68	異国の	
60	感染性の		69	リスク	
61	寄生の		70	利益	

2. つぎの各文が本文の内容と一致するものにはT(True)，一致しないものにはF(False)を，文末の（　）に記入しなさい．

（1）You can introduce an alien plant species without authorization by any legal entity if you are a scientist. (　)

（2）Natural migration also influences drastically on the speciation and consequently ecosystem in a short term. (　)

（3）The sixth extinction spasm is caused by the humans. (　)

（4）Tomato originated in South America. (　)

（5）Plant species with an allelopathic chemical substance, could be invasive to the newly released environment. (　)

3. つぎの日本語の各文を（　）の中の語を用いて英語の文にしなさい（必要があれば単語を適切な形に変換しなさい）．

（1）セイタカアワダチソウは，日本，中国やヨーロッパ諸国で，生態系に侵略し，在来植物へ危機をもたらした．(invade, threat)

（２）生物多様性の保存は，初等教育や社会啓蒙によって強化できる。(enforce, conservation, primary education)

（３）外国からの種の輸入は数種類の法律で規制されている。(regulate, import, alien)

（４）人間の行動が生物多様性に影響した結果は，遠い未来ではなく迅速に現れる。(consequence)

（５）一度絶滅した種を再生することは不可能であり，破壊された生態系は元に戻せない。(irreversible, extinct, regeneration)

Chapter 10　Invasive Species and Crisis in Biodiversity

4. つぎの各問いに英語で答えなさい。

（ 1 ）　What is the cause of the present sixth extinction spasm?

（ 2 ）　Would a strike of a huge meteorite cause an extinction spasm?

（ 3 ）　How can we avoid the influence to the biodiversity?

（ 4 ）　Has the movement of crop species been benefited humans?

（ 5 ）　Can all alien species be invasive?

5. つぎの日本語課題を英語で議論してください。

（ 1 ）特定外来種について皆さんはどのようなことを知っているでしょうか？

（ 2 ）すべての外来種は侵略種ではありませんが，海外から生物を新規に導入することはやっていいことでしょうか？

Chapter 11   Facing Environmental Degradation

Environment consists of inanimate beings and animate beings. Inanimate beings are such as atmosphere, sea, terrain, and their components like oxygen, carbon dioxide, water, soil, minerals etc. Animate beings are all living organisms such as moss, fern, tress, grasses, fungi, bacteria, insects, fish, birds, mammalians and humans. Inanimate and animate beings interact and they form up a variety of ecosystems leading to creation of biological diversity.

As pointed out on the biodiversity loss in the Chapter 10, now we are exposed to serious environmental degradation, which cause the sixth extinction spasm of biological diversity. Environmental degradation is well recognized by local residents and global communities as we are suffered from more unpredictable natural disasters, natural resources reduction and our own sake of daily activities.

Human activities have changed surely the global climates by deteriorating the environments. Forests are cut down without replacement of trees and other vegetation such as by slash-and-burn, and natural forest fires agitate and worsen the situation at the global level. These are serious damages to conservation of landscape and biological diversity.

The consequences are not only the environmental damages, but also people get suffered from deteriorated living conditions because resources are scarce with the losses and more occurrences of secondary disasters such as landslides due to the lack of appropriate vegetation. Also air, water and soil are more polluted or contaminated at

Note :
slash-and-burn「樹木の乱獲」

Chapter 11　Facing Environmental Degradation

many locations in the world due to over-exploitation of natural resources, and wastes and emission from large-scale industrial development and your modern life activities such as car driving and electricity uses.

We also require more food stuffs from natural conditions because of the boosting global population and also more demand for better food materials especially for fish and animal sources.

Natural fish populations are indiscriminately caught at sea and freshwater, and alien species such as tilapia and Nile parch are used for fish cultivation but often they invade deliberately introduced environments ending disturbance and destruction of the original ecosystem.

Meat products requires more primary materials and energy: one kilogram of fish meat needs 1 to 1.5 kg of feed, 2 to 2.5 kg for chicken, 5 kg for pork and more than 10 kg of food stuff for beef. The feed used for them are corn, soybean etc, which can be also used as food for human. The concerns are not only on the material diversion, but also on energy to be used in production of the feed, and on waste and emission that could come from the animals and processing them.

Imagine if these feed is used for human feeding instead of animals how much food, energy and efforts can be saved? Alternative debates can be raised that somebody should take more economic advantage to supply high value products in sparing the resources. It is true that commercial activity is common understanding and world prosperity depends on such a business interest at present.

We also cannot neglect the natural incidents. There are more

Notes :
tirapia「ティラピア，食用淡水魚」
Nile parch「ナイルパーチ（アカメの仲間）大型の食用淡水魚」

and harder disasters such as hurricanes, cyclones, earthquakes, volcano eruptions, tsunamis etc. We also have accumulative adverse effects by torrential rain, drought, soil erosion, salinization etc. These problems further create more fatal concerns: global warming and its consequences such as deforestation, desertification, melting of glaciers at the poles and high mountains, biodiversity decline etc. It is estimated that we are losing global forests equivalent to a size of one soccer ground at every second. Arable lands become barren at a rate of 150 to 200 million hectare every year in the world.

The average temperature of the earth will be two degree celcius higher in 2100. This would imply that more loss of trains with the increase of the sea level. Indeed, Tuvalu islands are now sinking into the sea. Coral reefs are also extinguishing at many coastal areas due to the temperature increase and with the invasive species such as starfish.

The loss of coral reef is not only on the sadness of the beautiful landscape, the coral reef species absorb the carbon dioxide and transform to oxygen by accommodating carbon to make the body of the coral reefs.

The global warming is attributed mainly to the emission of the carbon dioxide which are the consequence of modern human activities. Do you know that average Japanese produce one kilogram of CO_2 daily (**Figure 11.1**). So that a Japanese makes 365 kg per year. Imagine that you are making several times more CO_2 than your body weight per year.

Notes :
hurricanes, cyclones 「日本語では台風 (typhoon), 地域によって呼称が変る」
tsunami 「津波」
Tuvalu 「ツバル,南太平洋の島国」

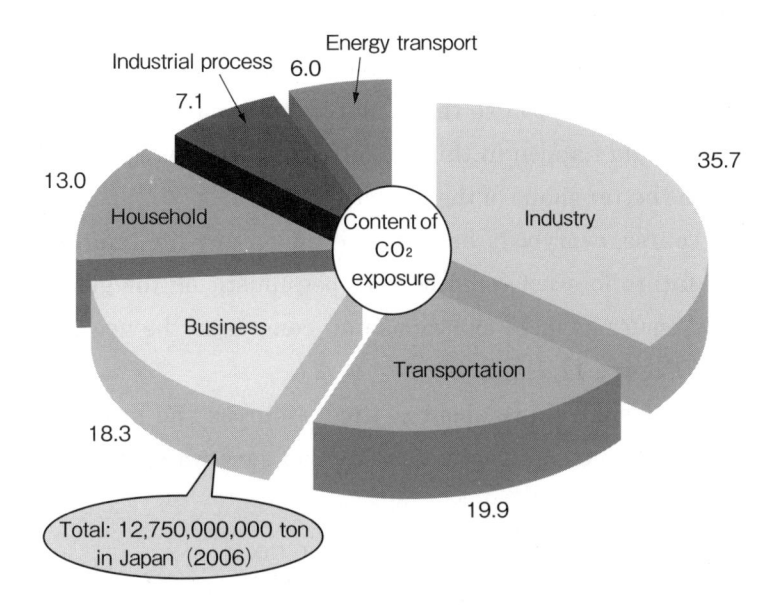

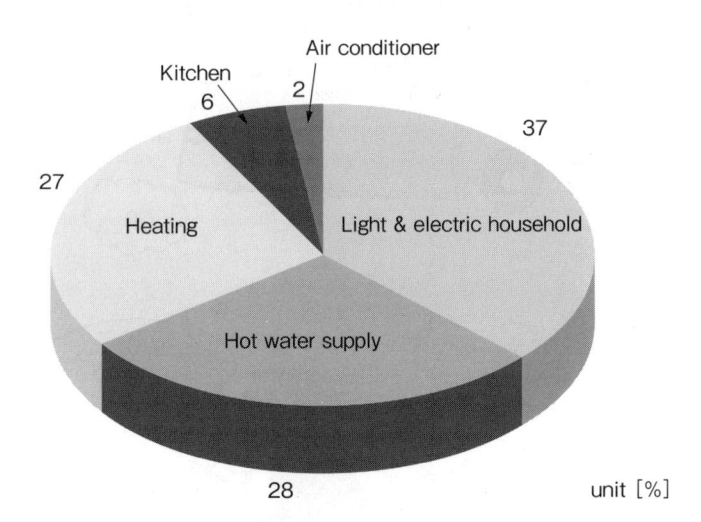

Chapter 11 Facing Environmental Degradation 97

Content of CO_2 exposure:
- Industrial process 7.1
- Energy transport 6.0
- Household 13.0
- Business 18.3
- Transportation 19.9
- Industry 35.7

Total: 12,750,000,000 ton in Japan (2006)

- Air conditioner 2
- Kitchen 6
- Light & electric household 37
- Hot water supply 28
- Heating 27

unit [%]

Figure 11.1 Enormous amount of CO_2 exposed and each person produces 1 kg of CO_2 one day
(modified from http://www.food-mileage.com/ and http://www.eic.or.jp/ecoterm/)

Many of people at developing countries do not have opportunity enjoying their life by spoiling the energy and resources. If they catch up with more chance to use the resources and make emission of the CO_2, what would happen in this planet? Unbalance in equity actually makes some better shape of the planet.

Of course, everybody has equal right on any life improvement, but close future forecast could be very pessimistic on the global environmental issues as more emissions now come from the newly emerging world (**Figure 11.2**).

Now we shall think about saving resources and protecting our environment by our daily activities? We can save water such by reus-

Figure 11.2 Modal shift is essential to stop global warming

ing the bath water for washing cloths and use rain for flushing toilet. Also avoid polluting water by eating up food and reduce use of the water for dish washing.

We can reserve electricity simply by reducing the uses of domestic electric equipments such as TV and also de-plugging the electric stuff which are unused. These efforts can reduce up to 9.4 % of the domestic uses of the electricity. Cool-biz and warm-biz make sense, and should we follow with the campaign. We also know that recycle, reuse and reduction of the use of the industrial materials. Bring your shopping bag "my bag" instead of using the plastic bags from the shops.

Why do we to try more intentionally on these things. Small actions could make a big and positive result.

Exercises

1. つぎの1～52の語に対応する英単語または英熟語を本文から選び出して書き込み，また発音しなさい（動詞は原形を記入しなさい）。

	Japanese	English		Japanese	English
1	劣化		6	大地	
2	構成する		7	生物	
3	非生物の		8	コケ	
4	生物の		9	シダ	
5	大気		10	カビ	

Note：

cool-biz and warm-biz「クールビズ（冷房の使用の制限），ウオームビズ（暖房の使用の制限），日本政府による省エネキャンペーン」

Unit C Environment

	Japanese	English		Japanese	English
11	細菌		32	無視する	
12	ほ乳類の		33	地震	
13	創成		34	火山	
14	予測しがたい		35	噴火	
15	天災		36	浸食	
16	劣化する		37	塩積化	
17	増長させる		38	森林破壊	
18	悪化させる		39	砂漠化	
19	景観		40	溶ける	
20	山崩れ		41	氷河	
21	混入して汚染する		42	極地	
22	汚染する		43	減少する	
23	排出		44	耕作可能な	
24	もの		45	不毛の	
25	増長させる		46	沈む	
26	非選択的に/無差別に		47	珊瑚礁(さんご)	
27	撹乱		48	海岸（の）	
28	破壊		49	悲観	
29	転用		50	不均衡	
30	加工する		51	予測	
31	節約する		52	悲観的	

2．つぎの各文が本文の内容と一致するものには T(True)，一致しないものには F(False) を，文末の（　）に記入しなさい．
（1）Animate things do not include viruses.（　）
（2）Earthquakes do not cause irreversible damage to the environments.（　）
（3）Our daily efforts do not help avoiding environmental degradation.（　）
（4）Switching-off and de-plugging can reduce the electricity uses.（　）
（5）Coral reefs absorb the carbon dioxide.（　）

3．つぎの日本語の各文を（　）の中の語を用いて英語の文にしなさい（必要があれば単語を適切な形に変換しなさい）．
（1）非生物環境要素と生物は相互にかかわり，多様な生態系を生み出し，その結果，生物多様性を創成する．(interact, form up, create, inanimate being)

（2）景観や生物多様性の保全に重篤な障害があった．(damage, landscape)

（3）ブルーギルのような侵略性の外来種は，在来の水生生物を駆逐する可能性があるだけでなく，ホテイアオイのように生態系全体を破壊する場合もある．(bluegill, risk, water lilies, extinct, invasive, destroy)

Unit C　Environment

（4）天災がいつ起こるか予測できないが，十分な予防措置は必要である。(predict, natural disasters, precautionary approach)

（5）日常生活で，どのように資源を節約し環境を保護できるのか考えてみよう。(save, protect, resources)

4．つぎの各問いに英語で答えなさい。
（1）Should we exterminate all invasive species form the planet?

（2）How can we reduce the emission of carbon dioxide at daily activities?

（3）Is it unfair that not all nations make policy and implementation of recycling?

(4) Japanese should not help other countries where the climate changes seriously affect as Japan itself has plenty of domestic problems at different aspects and also particularly on the environmental issues?

(5) How can we make our life more frugal and environmentally friendly?

5．つぎの日本語課題を英語で議論してください。
(1) 地球環境変動は，日常生活にも影響しているでしょうか？

(2) 海外で起こっている公害や環境問題は，他国のできごとで，われわれは関知しなくてもよいのでしょうか？

(3) 日々の健康維持と同じで，地球環境の健康改善は，個々の人間の努力が必要です。どのような日常行動が考えられるでしょうか？

Unit D Summary

Chapter 12 Environment and Ethics

It is always important and even ought to have ethical consideration and conduct in your life. Environmental aspect is the same, and indeed, environmental ethics should be of your knowledge and a part of your discipline to exercise environmental considerations as a citizen.

Environmental ethics is an emerging area which globally people should take into consideration into their life. Please look in the comprehensive references for your further interest, which are overviewed in the list of information at the end of this chapter.

Environmental ethics has been grown out from the fundamental philosophical issues on nature, and probably one of the oldest thoughts through the human history including the viewpoints out from religions and legends. The environmental ethics has influenced on different academic and practical disciplines. Even it has been exerting to the processes for making international laws and also to the content of domestic legal instruments. Because of the influences on the legal aspects, there have been also on practices such as sociology, economics and ethnology. Philosophy and theology also receive the feed-backs from the environmental ethics. Natural sciences are also enriched by the environmental ethical consideration such on the development of ecology and implementation on geography.

There is no argument that environmental ethics is the part of environmental philosophy; it contains properly components such as from aesthetics, theology and logics. However, it is not only as a branch of metaphysical issues, but also it includes the advices for the

Chapter 12 Environment and Ethics

practices in relationships of human beings with environments.

Philosophical points of the environment issues are: 1) sustainability, fairness and equity on the stakeholders of the environments. Then, who and what are the stakeholders? This is one of major arguments in the environmental ethics forum.

Neighboring residents or global citizens? Do we consider only human beings? Should we accommodate the animals? We must include all creatures as the stakeholders? Some deep ecologists contemplate the environment per se is the one! Due to the rapid development of genome and biotechnology research and development and with their social influences, even some conservative societies regard that a gene which is actually chemical property, could be as some component of a stakeholder?

Different people think different ways and it cannot be denied fully. Anyway, the most important issue is, whenever you think about the environmental issues, to have multidisciplinary approach and to examine different possibilities.

There are three principles on environmental ethics: Limit of natural resources and environment on the earth, duty and ethics to the unknown future generations, and right of the nature.

Aldo Leopold is regarded as the father of environmental ethics. He studied forestry and appreciated the nature both as respectful and mysterious as well as resources which helps human life at the beginning of 20 th century.

He served for United States Forest Service and he mostly did wildlife and game surveys throughout the U.S. later he become a professor at Univ. of Wisconsin-Madison. With his professional carrier and academic background as well as personal incline to the nature, he realized that the human activities really change and deteriorate the earth. He stated his thinking in his book "A Sand County Almanac", in which he proposed Land Ethic as a new ethical aspect. This is an advocate for the preservation of wildlife and wilderness areas.

You might have heard about a book, " Silent Spring" by Rachel Carson (1962). This classic writing raised the environmental concerns and indeed addressed real problems by the pollutions that destroy environments leading to extinction of wild life and causing troubles to human life including serious health issues such as cancer. She tried to ban synthetic pesticides which harmed the environments and human-beings, and become the pioneer of the environmental movement.

Ecocentrism, biocentrism and animal rights have come out by the movements of human activities more aggressive toward the nature and by the consequent paradigm shifts. In the following sections, each category shall be briefly discussed.

Anthropocentrism address the highest priority of human survival in the earth. Ecocentrism is contrast to anthropocentrism. Garret Hardin pointed the needs in serious consideration on the decreasing natural resources and deteriorating ecosystem.

He gave lessens on the needs of cohabitation of human and ecosystem by "The Tragedy of Commons, 1968" and "Ethical Implication of Carrying Capacity, 1977". With former work, he advocated that the earth is common resources for everybody and there is limitation of the benefit from the ecosystem.

The latter publication points out reality of global imbalance on the natural resources and wealth. Developed nations also have limitation to help developing countries on the poverty and living conditions, and if continuously the help is made, the wealthy societies also can be degradable and may end to a total destruction. Some degree of selfishness may also be needed rather than humanitarian issue. Callicot also followed the concept of the limitation of the anthropocentrism.

Biocentrism comes from respectful creed and philosophy to the life of the creatures on the earth. Schweitzer and Taylor would represent the thoughts. Nash states the recognition of the right of the nature in his 1980's publication, "The Value of Wilderness". He addressed that human cannot control all the nature and nature per se has the influence on the direction of its presence.

Singer and Regan proposed and made actions on the animal freedom and animal right, respectively. These activities now make cardinal consideration and legal systems to care and protect from abusing animals such as by animal experiments and livestock uses when they are slaugthered. Also we must contemplate on who decide the survival of monkeys and deer which disturb agriculture in Japan.

Can we shoot the crows because they make messes with garbage? Immigrants brought rabbits into Australia for their sake as food, now rabbits become wild and cause environmental degradation. Then, rabbits can be exterminated? Now we have animal ethics as one disciple of the environmental ethics and also of biethics. Human is not the only presence on the earth and we should not to dictate in the planet.

Different regions of the world have been considering similar aspects under their religious and cultural backgrounds. Asians also have own thought on the environmental ethics throughout their history. Particularly on Japanese, Bushido had been changed from fighting spirits to the discipline for living, the environmental ethical thought have been accommodated.

As environmental concerns become global recognition, many international discussion have been made such as UN Brundtland Com-

Notes :
UN Brundtland Commission「環境持続性に関するブラントランド委員会, http://www.un-documents.net/a42r187.htm」

mission (1987), UNCED (1992) and WSSD (2002). As the consequences, international laws have been agreed as common understanding to implement the environmental conservation and sustainable uses.

The examples are such as Kyoto Protocol (1994) on the emission of carbon dioxide and other substances from human activities to avoid global warming under UNFCCC; Convention on Biological Diversity (1993); and now IPCC gets a lot of recognition due to the real global problem and appreciation of the efforts indicated by the Noble Prize Laureate to IPCC in 2007. During the processes of such international establishment, positive conceptual influence was accommodated from the environmental ethics.

Then, how do you think shallow ecology which puts more priority for humans and deep ecology that critically discuss on the ecocentrism, biocentrism and animal rights? Do you admit the word, sustainable development, globalization of economical growth by sacrificing the environment, on the other hand, we must face the pollution and full of adverse effect of our life to the earth, and can we be selfish to seek for our life better?

Please keep in mind four aspects when you think and conduct environmental issues: environmental sustainability, fairness, equity, consideration and dedication to all stakeholders.

Notes:

UNCED：UN Conference on Environment and Development「地球サミット（リオサミット），www.un.org/geninfo/bp/enviro.html」

WSSD：World Summit on Sustainable Development「世界持続的発展サミット（ヨハネスブルグサミット），www.un.org/events/wssd/」

Convention on Biological Diversity「生物多様性条約，http://www.cbd.int/default.shtml」

Exercises

1. つぎの 1 ～ 75 の語に対応する英単語または英熟語を本文から選び出して書き込み，また発音しなさい（動詞は原形を記入しなさい）。

	Japanese	English		Japanese	English
1	倫理（学）		19	集会	
2	規律/修養		20	住民	
3	出現する		21	受容する	
4	包括的な/わかりよい		22	生態学者	
5	哲学		23	ゲノム	
6	社会学		24	バイオテクノロジー	
7	論理学		25	影響する	
8	経済学		26	保守的な	
9	生態学		27	化学の	
10	地理学		28	特性	
11	美学		29	要素	
12	部門/枝		30	学際的な	
13	調査		31	取組み	
14	持続性		32	義務	
15	公正		33	世代	
16	衡平		34	林学	
17	関係者/ステークホルダー		35	尊敬する	
18	議論		36	神秘な	

Unit D Summary

	Japanese	English		Japanese	English
37	資源		57	優先	
38	野生動物		58	指摘する	
39	履歴		59	共存	
40	悟る		60	限界	
41	傾注する		61	意図	
42	劣化する		62	利益	
43	状態		63	出版物	
44	唱導する		64	現実	
45	保存		65	富	
46	野生生物		66	貧困	
47	荒れ野/自然		67	人道的な	
48	喚起する		68	分解する	
49	重篤な/真剣な		69	利己的な	
50	合成の		70	乱用する/虐待する	
51	殺虫剤		71	生命倫理学	
52	絶滅		72	絶命させる	
53	草分け		73	排出	
54	生態中心主義		74	概念的な	
55	生命中心主義		75	献身	
56	人間中心主義				

2. つぎの各文が本文の内容と一致するものには T(True), 一致しないものには F(False) を, 文末の (　) に記入しなさい。

(1) Environmental ethics is an emerging area which globally people should take into consideration into their life. (　)

(2) Environmental ethics is the part of environmental philosophy. (　)

(3) Environmental ethics is simple discipline and does not include environmental aesthetics, environmental theology, and all the other branches of philosophical investigation. (　)

(4) During the processes of such international law establishment, positive conceptual influence was accommodated from the environmental ethics. (　)

(5) We need thinking and conducting environmental issues without environmental sustainability, fairness, equity, consideration and dedication to all stakeholders. (　)

3. つぎの日本語の各文を (　) の中の語を用いて英語の文にしなさい（必要があれば単語を適切な形に変換しなさい）。

(1) 新しい環境問題が次々と生じてくることを, われわれは認知しなければならない。(emerge, realize)

(2) レオポルドは, 野生動物と自然を崇拝し, 環境倫理学を唱導した。(advocate, wildlife and wilderness)

（3）適正な殺虫剤の使用は，必ずしも重篤な環境負荷になるとは限らないことを喚起すべきである。(raise, serious, pesticide, hazard)

（4）彼女は生態中心主義の立場から環境保護に傾注した。(ecocentrism, incline, view point)

（5）環境問題への対処には，優先度の検討が必要であることを指摘すべきである。(priority, point)

4．つぎの各問いに英語で答えなさい。
（1）Is the cohabitation of human and other creatures possible?

（2）Can you take deep-ecology thoughts for practicing sustainable development?

（3）What is the implication of the international laws such as FCCC?

（ 4 ） Can you share your resources with people at developing world?

（ 5 ） Review the key components of environmental ethics.

5．つぎの日本語課題を英語で議論してください。

（ 1 ）あなたは，人間中心主義とディープエコロジーのいずれについて受容できますか？ 中庸ですか？

（ 2 ）家畜などの動物の権利は，どこまで認めることができるでしょうか？

（ 3 ）バイオテクノロジーの進展によって，生命の操作が多様な科学分野で行われつつあります。科学の利便性の一方，倫理，宗教，個別の心情などから，どのようにこれを受け止めているでしょうか？

Chapter 13 How Would be Our Efforts to Daily Life to be Environmentally Friendly?

Do you know that our modernized life waste a lot of resources? Do you turn off the light switches at home when you are not there? You believe that you never wasted food in your refrigerator? Can we say that we are really saving resources? Even before we consume a lot of food stuff, energy and materials for living, they are spoiled or lost. It should be revisited on the 3 R as in Chapter 6 on the recycle, reuses and reduction.

For example, seventy percent of energy produced at the electricity plants by nuclear or thermal power generation, could be lost without uses, and these wastes also associate with global warming. It is striking that only thirty percent of such generated energy are to be used as electricity even at the developed world. On the other hand, efforts have been made to reuse the auxiliary energy produced while electricity is generated: this is called co-generation system.

The room heating and hot water supply can be supported by the unused heat from such electricity generation. Thus, the primary energy generated from such plants can fully be utilized. Also with the use of such heating sources, and SO_2 can be reduced up to 50 %, compared with generating heat specifically for warming rooms and water. Furthermore, it is the time to see sustainability of energy supply and reduce CO_2 by using alternative electricity generations such as wind (**Figure 13.1**).

Besides the efficient uses of energy, saving energy and reducing uses are also cardinal: for an example, many of offices and schools in Germany turn down the use of electricity for half an hour or so every day, even with this reduces the generation of electricity and can divert the unused amount to meet with the skyrocketing demands.

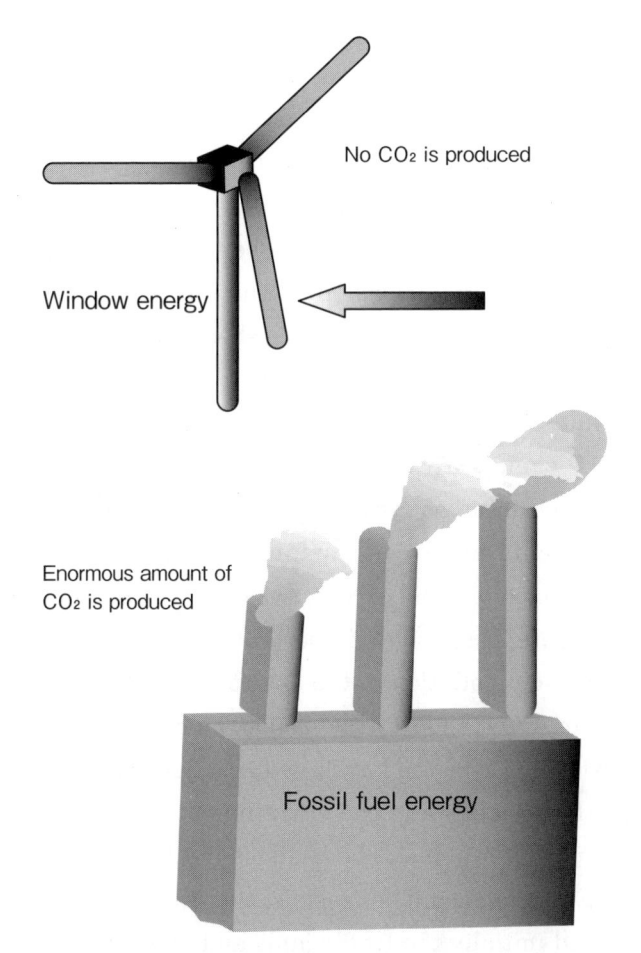

Chapter 13 How Would be Our Efforts to Daily Life to be Environmentally Friendly?   115

Figure 13.1 The change of energy resouces

A heliotrope, is one of innovative ways to use the sunlight. Please do not confuse with heliotrope plant (*Heliotropium peruvianum*). This is a sunlight-based electricity-generating house which rotates 360 degree accordingly to the direction of sunlight to receive

---

Note :

heliotrope「キダチルリソウ（木立ち瑠璃草）。香水で有名」

the energy from the sun. Rain water conservation for house-keeping, compost toilets and replacement of bulb to fluorescent light sources, are also small efforts, but they are not negligible as an accumulated amount for the natural and energy resources uses.

Legal and social system changes are also aspects to consider. End-retailers of disposal materials can be taxed when they sell such items as they must be burnt or dealt as industrial wastes. Public events such as festivals at public parks, should not allow garbage and unrecyclable wastes.

Recycling could produce a lot of opportunities by direct reproduction of the same entities such as cans and papers, reusing the same items by cleaning as glass bottles, or converting them to other materials. Also enterprises with manufacturing shall encourage the work hour restricted to avoid excess uses of energy and materials.

Furthermore, it is always important to maintain a high level of compliance to avoid pollution of the environments by reducing industrial wastes. An ISO standard (ISO 26000) on corporate social responsibility (CSR) will also be the key issue to conduct the corporation contribution to the societies and the environmental consideration is the one of the major components to be included in CSR.

Many European countries restrict the use of private cars in city centers to avoid initially the traffic jams and now because of environmental concerns. Ride-sharing and car-ownership sharing are common, and encouragement to use public transportation system is another option.

As for waste disposal, even an eelworm can helps garbage han-

Notes:

ISO : International Organization for Standardization「世界標準化機構」

corporate social responsibility (CSR)「企業の社会貢献」

dling. Small amount of raw food waste can be eaten up by the creature at your backyard garden and the consequence produces composts which are very effective to keep the soil productivity.

Then, how about should we look for reproducible and recyclable entities instead of consuming dwindling natural resources? We had discussion in the previous chapter on the bioenergy and biomaterials, and it is hope for survival of human beings. There may be over-expectation in term of time frame for materializing the application of technology associated for a close future uses.

However, sooner or later, we should be able to have more bio-based materials and energy to make our planet sustainable besides balancing with the systematic uses and conservation of the natural resources.

As we already have discussed in the previous chapters, we need food for feeding global population while keeping natural resources and environments over-exploited.

Before improving the methodology for production to fulfill demands of consumers, the post-production handling such as post-harvest technology is more important for immediate and cost-effective elaboration for reducing such losses, meanwhile modern science and technology shall be lined up for next options.

There are many ideas and implementation activities for your surroundings. Please always think how your small action can help and save your environment and own future with limited natural resources. Each of us must make own effort without exceeding the capacity what we can do, and it is also cardinal to recognize that only with public engagement at all communities in this planet, the efforts can make it through gradually. We must believe that a small thing by each individual can make it possible to the global alleviation of so many tasks on the environmental issues.

Unit D　Summary

Exercises

1. つぎの 1 ～ 43 の語に対応する英単語または英熟語を本文から選び出して書き込み，また発音しなさい（動詞は原形を記入しなさい）。

	Japanese	English		Japanese	English
1	廃棄物/むだにする		19	交換	
2	現代化した		20	蓄積する	
3	消す		21	法的な	
4	点燈する		22	販売業（者）	
5	消費する		23	燃やす	
6	省力する		24	処置する	
7	むだにする		25	処理する	
8	火力発電		26	処理	
9	施設/工場		27	ゴミ	
10	核の		28	リサイクル可能な	
11	予備の		29	リサイクル	
12	コジェネレーション		30	再生	
13	転用する		31	再利用	
14	飛躍的な		32	製造	
15	削減する		33	勧奨する	
16	回転する		34	制限する	
17	家事		35	遵守	
18	堆肥		36	貢献する	

Chapter 13　How Would be Our Efforts to Daily Life to be Environmentally Friendly?　**119**

	Japanese	English		Japanese	English
37	交通渋滞		41	収穫	
38	生産性		42	向上	
39	実行化する		43	実施	
40	開拓する				

2．つぎの各文が本文の内容と一致するものにはT(True)，一致しないものにはF(False)を，文末の（　　）に記入しなさい。

（ 1 ）It does not help to maintain natural resources by a small effort. (　　)

（ 2 ）Present electricity generation systems could have more efficiency in energy diversion. (　　)

（ 3 ）New findings on science and technology can only assist solving the present global warming. (　　)

（ 4 ）Use of eelworms can be one of sustainable ways to dispose domestic garbage. (　　)

（ 5 ）Europeans do not make any regulation on car uses. (　　)

3．つぎの日本語の各文を（　　）の中の語を用いて英語の文にしなさい（必要があれば単語を適切な形に変換しなさい）。

（ 1 ）お金が掛からなくても天然資源をむだ使いすることは正しくない。(abuse, natural resources)

（ 2 ）食料は，むだにしないように適正な計画により輸入および供給すべきである。(spoil, import, supply)

Unit D　Summary

（3）原子力発電は，効率が悪いわけではない。(nuclear, generate)

（4）天然資源の持続的利用は，リサイクル，再利用および過剰な利用の削減に依存するところが大きい。(depend, sustainable use, excess)

（5）日常生活における省力の努力は高く勧奨されている。(encourage, save)

4．つぎの各問いに英語で答えなさい。
（1）Is the co-generation of energy helpful for cost-reduction?

（2）How can you make yourself contributed for environment protection?

(3) Can you make a manufacturing business under so much of legal and social rules on the environment and natural resources?

(4) How do you think ride-sharing?

(5) How do you contribute to your society after reading this chapter?

5. つぎの日本語課題を英語で議論してください。

（ 1 ）ガソリンの高騰があります。そして，車は，二酸化炭素の排出原因になっています。単価が高くても電気自動車やハイブリッドカーを購入するつもりはありますか？ あるいは，車に乗ることを止めるほうが環境に優しいことになりますが，どのような選択肢を考えますか？

（ 2 ）省エネのために日常生活でどのような努力をしているでしょうか？ 公共施設などでのむだ使いは特に著しいですが，学校などでの努力はあるでしょうか？

Chapter 14 Epilogue: Challenges at Tsukuba Science City

Here an introduction is made as an epilogue of this book on a regional effort with innovation for future.

University of Tsukuba, has proposed in December, 2007 "Tsukuba Eco-City Initiative" as the core plan on coordination and collaboration at Tsukuba Science City, Ibaraki, Japan. Tsukuba Science City consists of more than two hundred research and technology institutions under different sectors and high technology industry, and they can be both the cause of environmental concerns and also instruments for the remediation.

Key organizations agreed to participate and implement the proposed plans.

Those representing institutions are: Local governments of Tsukuba City and Ibaraki Prefecture, Advanced Institute on Science and Technology (AIST), National Institute of Environment (NIE), National Institute for Materials Science(NIMS) and University of Tsukuba (UT).

The basic concepts are: 1. The themes of the activities shall be the common agenda at the Science City; 2. Encouragements of the participation of various organizations with their specialty; 3. UT shall have the institutional commitment as the core of the Initiative; 4. The plan and activities should meet with the Science and Technology Policy of the Japanese government; and 5. The plan and activities also should meet with the third stage development plan of the Tsukuba City.

The plan is made of four subprograms.
 1. Tsukuba Environment-Energy-Economy (3E) Program
   The Program is targeted for development of technology and

systems on energy-efficiency and clean energy generation for the energy efficient city planning.
2. Environmental Conservation Model Program
The program is aimed for creation and trials on the environmental technology and system for the local model society with resources recycling and coexistence of different activities.
3. Education, Culture and Health Model Program
The component is oriented for the development of the residential system with the respects to the health, education, culture and local cultural heritages.
4. Rejuvenation of City Function Model Program
City Planning is discussed and formed with participatory communications for safety in living, low pollutions, energy-efficient transport system and information system on disasters.

With the planning and implementation, Tsukuba Science City shall be restructured as the national and even global model on the environmentally-friendly city with strong science & technology and science education functions. As the initiation of the four programs, the 3E forum is proceeded, and other three will follow as the participation will be enforced.

つくば 3E フォーラム事務局

筑波大学大学院生命環境科学研究科内
〒 305-8572　茨城県つくば市天王台 1 丁目 1-1
TEL&FAX: 029-853-4301
E-mail : secretariat-3ef@sakura.cc.tsukuba.ac.jp

つくば 3E フォーラム
Tsukuba Environment-Energy-Economy Forum

The 3E forum has an ambitious target: Tsukuba Science City shall reduce its exhaustion of carbon dioxide to the 50 % of the present level by Year 2030. And this efforts shall be spread over the nation in 2030 to 2050 as the representation.

The Science City shall make the integrative efforts with coordination and collaborations by various institutions with distinguished science and technology capacity as the team in order to recommend and demonstrate the essential technology and system to achieve such a goal to make the environmentally-friendly city.

Such an integration of the technology and knowledge should be the model for creating low carbon emission societies and the model shall be introduced to global communities.

Appendix

Importance of Conservation and Sustainable Uses of Natural Resources: Biological Diversity and International Challenge towards 2010 (http://www.bioversityinternational.org/Themes/Conservation_and_Use/index.asp)

Japan will host the implementation meeting, the COP-10 of the Convention on Biological Diversity in 2010. CBD was established in 1992, and it contains typical examples on the sustainability issues with environments and natural resources. There will be increasing interests on the CBD and biodiversity in Japan and elsewhere and the key elements are addressed in this appendix. This text is quoted from the information by Bioversity International, which is one of very active international scientific research organizations for the promotion on the conservation and sustainable uses of biodiversity.

The loss of biodiversity poses a serious threat to agriculture and the livelihoods of millions of people. Conserving biodiversity and using it wisely is a global imperative. Biodiversity provides the foundation for our agricultural systems. It provides the sources of traits to improve yield, quality, resistance to pests and diseases and adapt to changing environmental conditions, such as global warming.

Biodiversity is also a direct source of food for many people and is a essential part of our life support system. Without it our ecosystems, the planet's entire biosphere, cannot function. There are different approaches to conserving biodiversity and different ways of using genetic resources. The four items are the components on the conservation; I. On farm management; II. *In situ* conservation; III. *Ex situ* conservation; and IV. Complementary conservation.

I. On farm management

On farm management involves the maintenance of crop species on farm or in home gardens. The effectiveness of strategies to maintain and use crop or livestock diversity on farms depends on the extent to which local varieties continue to meet the needs of farmers and communities. The approach taken needs to be one that is embedded in the community and reflects its values and concerns.

Many plant genetic resources, especially those of minor crops, are managed as part of agricultural production systems. This type of biodiversity conservation has been termed 'conservation through use'. There are important reasons for supporting on farm maintenance of crop and livestock diversity:

1) It ensures that the ongoing processes of evolution and adaptation of crops to their environments.
2) It allows for the continued selection of superior material by farmers that meets their needs and preferences.
3) It helps preserve indigenous knowledge, strengthens local institutions and promotes farmers' participation in national biodiversity conservation programmes.
4) It provides a necessary backup to genebank collection.
5) It provides natural laboratories for agricultural re-

search.

Internaitonal organizations such as Bioversity International focuses on exploring when where and why traditional varieties are maintained who maintains them and how this is done. In this way we try and identify options that can support conservation on farm and in home gardens.

Managing genetic resources on farms concerns the entire ecosystem, including cultivated crops, forages and agroforestry species, as well as their wild and weedy relatives that may be growing in nearby areas.

Conserving the processes of evolution and adaptation:

The conservation and use of agrobiodiversity at all levels within local environments helps ensure that the ongoing processes of evolution and adaptation of crops to their environments are maintained within farming systems. This benefit is central to *in situ* management of genetic resources, as it is based on conserving and using not only existing germplasm but also the conditions that allow for the development of new germplasm.

Conserving and using diversity at different levels:

In its maintenance of farming systems, on-farm conservation applies the principle of conservation and use to all three levels of biodiversity: ecosystem, species and genetic (intraspecific) diversity. In conserving the structure of the agroecosystem, with its different niches and the interactions among them, the evolutionary processes and environmental pressures that affect genetic diversity are maintained and this contributes to the overall health of the local environment.

Integrating farmers into a national plant genetic resources system:

Farmers are likely to know the nature and extent of local crop resources better than anyone through their daily interactions with the diversity in their fields. Given their expertise, incorporation of farmers into the national genetic resources system can help create productive partnerships for all involved. This integration can happen in several ways, including:

1) Seeing farmers as partners in the maintenance of selected germplasm
2) Establishing a national dialogue on biodiversity conservation, sustainable use and equitable benefit-sharing between farmers, genebanks and other partners
3) Assisting the exchange of information with and among farmers from different sites and projects
4) Farmers visiting genebanks or seeing demonstrations by genebanks
5) Developing systems to make genebank material more easily accessible to farmers.

Improving the livelihoods of resource-poor farmers:

In situ on farm genetic resource management programmes also have significant potential to improve the livelihoods of farmers at the local level. On-farm conservation and use programmes can be combined with local infrastructure development or the increased access for farmers to useful germplasm held in national genebanks.

Farmers benefit from the continued agricultural diversity and ecosystem health that these programmes support. Local crop resources can be the basis for initiatives to increase crop production or secure new marketing opportunities. By building development efforts on local resources and through the empowerment of farming communities, they can lead to sustainable livelihood improvement. Resource-poor farmers, in particular, may benefit if development initiatives are not based on external inputs that may be costly or inappropriate for marginal agroecosystems.

Maintaining or increasing farmers control and access over genetic resources:
On-farm conservation and use also serves to empower farmers to control the genetic resources in their fields. On-farm genetic resource management recognizes farmers and communities as the curators of local biodiversity and the traditional knowledge to which it is linked. In turn, farmers are more likely to reap any benefits that arise from the genetic material they are managing.

Managing stress and change:
Managing agroecosytems and the biodiversity they contain is essential for human health and nutrition and for the continued availability of food and other agricultural products.

Crop diversity can be used as a resource to mediate potential stresses of the surrounding environment. A crop population with a diverse genetic makeup may have a lower risk of being entirely lost to any particular stress, such as temperature extremes, droughts, floods, pests, and other environmental variables. Crops with different planting times and times to maturity give the farmer the option to plant and harvest crops at multiple points in the season to guard against total crop loss to environmental threats.

Farmers shape the degree and distribution of genetic diversity in their crops both directly, through selection, and indirectly, through management of different agroecosystem components. For many farmers in developing countries, the availability of adaptive varieties for particular micro-niches may be one of the few resources available to increase or maintain production on his or her field.

Seed systems and diversity maintenance :
Each year farmers decide how much seed to plant and where that seed comes from. These seeds may come through formal and informal systems, and may contain new combinations of genes that result from hybridization and introgression between wild and cultivated plants or among cultivated varieties.

Whether new combinations of genes or new seed types are maintained, with the resultant development of populations with new characteristics depends on farmer management and access. Rather than being passive recipients of seed from the formal sector (government, extension agencies, seed companies), farmers participate in dynamic networks of seed exchange and development.

On farm biodiversity management programmes that support strong seed supply systems can foster increased use of diversity while fulfilling certain types of farmer seed demand. Strong seed supply systems enable farmers to maintain a high level of diversity over time, despite losses of seed stock, bottlenecks, and other regular or unanticipated losses of crops genetic diversity.

Research on seed systems has shown that:
1) a very large percentage of the seeds used by farmers in developing countries is acquired by informal means such as local markets, friends and relatives
2) farmers engage in the innovation of new varieties by actively seeking our new seed or saving seed of plants that display new traits in their fields
3) farmers retain diversity according to environmental conditions, market demands, culinary and aesthetic preferences, and social factors like religion and prestige
4) farmers blend modern and traditional varieties.

Seed systems may shape crop genetic diversity and that seed systems can act as linkages between very distinct populations.

Ecosystem services:
One commonly cited justification for maintaining plant genetic diversity in agricultural production systems generally, and in farmers' fields in particular, is its ability to buffer or limit

disease and pest epidemics. When many farmers sow varieties that carry the same genetic mechanism of resistance to a plant disease, the crop is vulnerable to epidemics. The complexity of crop-pest interactions in agroecosystems is compounded by their seasonal or annual variability, particularly in stressful environments of extreme temperatures and unpredictable rainfall.

Combating epidemics once they occur can be costly to society both in terms of garnering the resources necessary to control them and the yield losses incurred, and especially so in developing countries.

By conserving and harnessing ecosystem biodiversity, on farm genetic resource management programmes can provide other valuable services to farmers. For example, the value of having a diversity of pollinators, such as bees, butterflies, hummingbirds and even bats is immense. Plus, soil biodiversity helps keep farmers' field fertile.

Linking wild and cultivated systems:
On farm biodiversity management programmes can serve to increase out knowledge about the links between wild crop species and cultivate ones.Whether new combinations of genes that result from hybridization and introgression between wild and cultivated crops are maintained, with the resultant development of populations with new characteristics, depends on natural selection, and in the case of crops, on human selection.

While many cases of deliberate introgression of desirable traits into crop cultivars as part of breeding programs are known, the extent and significance of natural or farmer assisted introgression is uncertain. A range techniques has been used to document natural hybridization and introgression of agricultural crops and their wild relatives in many crops including maize, wheat, barley, oats, pearl millet, foxtail millet, quinoa, hops, hemp, potato, casava, common bean, cowpea, pigeon pea, carrots, squash, tomato, radish, lettuce, chili, beets, sunflower, cabbage, and raspberries. However, the majority of these studies are based on morphological characters, and few have investigated the frequency with which such new types are produced and retained in natural and agroecosystems for farmer selection.

Even more limited is information on the role of farmers in recognizing and selecting new genetic variation from the natural introgression of crops with their wild relatives, and the impact, once selected, of these new genetic combinations on the crop diversity.

II. *In situ* conservation

In situ (=on-site) conservation and use refers to the maintenance and use of wild plant populations in the habitats where they naturally occur and have evolved without the help of human beings.

The wild populations regenerate naturally, and are dispersed naturally by wild animals, winds and in water courses.

There exists an intricate relationship, often interdependence, between the different species and other components of the environment (such their pests and diseases) in which they occur. The evolution is purely driven by environmental pressures and any changes in one component affect the other. Provided that changes are not too drastic, this dynamic co-evolution leads to greater diversity and better adapted germplasm.

The conservation of the forests and other wild plant species is often carried out through, but not limited to, the designation of protected areas such as national parks and nature reserves. Bioversity's work focuses on the maintenance of the genetic dimension of the wild species, especially on forest and crop wild relatives in the protected areas and beyond to ensure that their wild populations are of sufficient diversity to allow them to adapt to the changing environments and in particular to climate change.

III. *Ex situ* conservation

Ex situ (= off-site) conservation of germplasm takes place outside the natural habitat or outside the production system, in facilities specifically created for this purpose. Depending on the type of species to be conserved, different *ex situ* conservation methods may be used.

The importance of genebanks has increased significantly over the last several decades. As commercial agriculture has expanded many farming systems that preserved local agricultural diversity have been transformed and local varieties have been lost. Controlling powerful social and economic forces so that they do not result in genetic erosion is often not possible, certainly not in the short term. As a result, genebanks often represent the only option for conserving biodiversity.

However, *ex situ* conservation is not just about conserving biodiversity for its own sake. The main purpose of the collections is to ensure agricultural growth and keep our options open for innovation. In genebanks, biodiversity is managed so that breeders, farmers and researcher can use it in their work.

To make the genetic resources useful to farmers, breeders and researchers, genebank managers must carefully document the collected materials, make the information available and establish a transparent and safe system for its distribution. They should take all the steps to make the material they conserve, including germplasm enhancement, is used by breeders and other researchers for agricultural development.

IV. Complementary conservation

A complementary conservation strategy can be defined as "the combination of different conservation actions, which together lead to an optimum sustainable use of genetic diversity existing in a target genepool, in the present and future". We should not forget that the main objective in any plant genetic resources (PGR) conservation programme is to maintain the highest possible level of genetic variability present across the genepool of a given species or crop both in its natural range and in a germplasm collection.

The various conservation approaches discussed above have distinct advantages and disadvantages, but the most effective conservation system should incorporate elements of both. They need to be used in a complementary manner. Nevertheless, in a number of cases it is not clear how decisions as to optimally balance the different methods available can be made how such a strategy would work. More research is required.

A complementary conservation strategy involves striking the right balance between different methods employed. It depends on the species being conserved, the local infrastructure and human resources, the number of accessions in a given collection, its geographic site and intended use of the conserved germplasm.

Such a strategy does not advocate a particular method, simply because the method is available, but because it is the most appropriate one under the given conditions. A good complementary conservation strategy does not categorize crops or species into definitive classes. It is dynamic, and lends itself to meet the challenges of changes that are occurring in the field of genetic resources as it is open to new technologies and new needs.

Importance of participatory efforts

The relationship between the conservation of genetic resources and their use is an intimate one. On-farm, *in situ* and community-level conservation and use have become integral to the conservation of plant genetic resources. Because of this, biodiversity research must take into account social and cultural factors, including decision-making patterns, local institutions, indigenous knowledge and value systems.

Community-based biodiversity management is a participatory approach to strengthen the capacity of local institutions and farmers for managing biodiversity for social, economic and environmental benefits.

The rationale of community biodiversity management is that the local community institutions or individuals with in a community can be strengthened through the management of community based knowledge systems to identify, conserve, manage, add value, and exchange on-farm local diversity.

Community-based biodiversity management is embedded in existing social structures and local institutions ranging from families to markets. Local systems of classification of crop and species diversity reflect socio-cultural perspectives for recognizing and using genetic diversity and its functional attributes. Locally managed seed systems are among the most important institutions for agricultural biodiversity management. Seed diversity and its associated knowledge is regulated by a set of specific rights, responsibilities and division of labor, often related to gender and age.

Social networks play a key role in determining access to seed and information. Relationships of trust and affection within the extended family or beyond are fundamental to the decision-making process, while norms, laws, rules, procedures, traditions, customs and practices influence the choice of individuals. All of which affect the movement of germplasm among households, villages and over larger geographic areas.

Maximizing the contribution of agricultural biodiversity to sustainable livelihoods may involve strengthening human and social capital in ways that support the management of the natural capital, including plant genetic resources. Biodiversity is an important but often undervalued asset available to resource-poor farmers for managing vulnerability, uncertainties, shocks and stresses. For this reason, issues relating to access to and control of local genetic resources are critical for the optimal management of biodiversity on farm.

© Bioversity International 2008

References

Chapter 1
参考情報
1）神門善久：日本の食と農，NTT 出版（2006）
2）JA グループ環境推進協議会編：食料と地球環境，家の光協会（1999）
3）大塚　茂：アジアをめざす飽食ニッポン，家の光協会（2005）
4）青沼陽一郎：食料植民地ニッポン，小学館（2008）

Chapter 2
参考情報
1）千葉　保 監，高橋由美子 絵：コンビニ弁当 16 万キロの旅，太郎次郎社エディタス（2005）
2）（独）農林水産政策研究所
http://homepage2.nifty.com/shokuiku/subspecial0412.htm (2008)

Chapter 3
参考情報
1）食品成分：
http://fooddb.jp/ (2008)
2）健康と栄養：
http://www.nih.go.jp/eiken/ (2008)
http://fnic.nal.usda.gov/ (2008)
http://mmh.banyu.co.jp/mmhe2j/sec12/ch152/ch152g.html (2008)

Chapter 4
参考情報
1）夏目書房編集部編：「買ってはいけない」は買ってはいけない，夏目書房（1999）
2）UTAN「驚異の科学シリーズ」編：今「輸入食品」が危ない，学研（1993）
3）川岸宏和：ビジュアル図鑑，食品工場のしくみ，同文館出版（2005）
4）鷲　一雄：またあるあるにダマされた，三才ブックス（2006）
5）週間金曜日編集部編：買ってはいけない，(株)金曜日（1999）
6）日本のテレビ番組がデータねつ造を認める，Nature DIGEST（2004 年 4 号）
7）橋本直樹：食品不安—安全と安心の境界，日本放送出版協会（2007）
8）松永和紀：食品の安全学，家の光協会（2005）
9）松永和紀：踊る「食の安全」，家の光協会（2006）
10）三好基晴：ウソが 9 割 健康 TV，リヨン社（2007）

11) http://www.mhlw.go.jp/topics/bukyoku/iyaku/syoku-anzen/index.html (2008)
12) http://www.hfnet.nih.go.jp/archive/c257471-2 (2008)
13) http://www.jhnfa.org/ (2008)
14) http://www.maff.go.jp/j/jas/index.html (2008)

Chapter 5
参考情報
1) 林　俊郎：シリーズ地球と人間の環境を考える 7, 水と健康, 日本評論社（2004）
2) Robin Clarke, Jannet King 著, 沖　大幹監訳, 沖　明訳：水の世界地図, 丸善（2006）
3) 国際調査ジャーナリスト協会著, 佐久間智子訳：世界の＜水＞が支配される, 作品社（2004）
4) 日高敏隆, 総合地球環境学研究所編：子供たちに語るこれからの地球, 講談社（2006）
5) 東京商工会議所編：環境社会検定試験 eco 検定公式テキスト, 日本能率協会マネジメントセンター（2006）
6) Stockholm Environmental Institute, http://www.sei.se/index.php?section=water (2008)
7) http://internationalrivers.org/ (2008)
8) http://www.iwmi.cgiar.org/ (2008)
9) http://www.gwpforum.org (2008)
10) http://www.un.org/millenniumgoals/ (2008)

Chapter 6
参考情報
1) 高月　紘：ゴミ問題とライフスタイル, 日本評論社（2004）
2) 成美堂出版編集部編, 左巻建男著：中学生の環境とリサイクル自由研究, 成美堂出版（2006）
3) 東京商工会議所編：環境社会検定試験 eco 検定公式テキスト, 日本能率協会マネジメントセンター（2006）
4) 田中章義編：地球では 1 秒間にサッカー場一面分の緑が消えている, マガジンハウス（2004）

Chapter 7
参考情報
1) 大聖泰弘, 三井物産(株)編：バイオエタノール最前線, 工業調査会（2004）
2) 松村正利・サンケアフューエルス(株)編：バイオディーゼル最前線, 工業調査会（2006）
3) 経済産業省編：エネルギー白書 2006 年度版, ぎょうせい（2006）

4）クリストファー・フレイヴィン編著：地球白書2006-2007，ワールドウォッチジャパン（2006）
5）N.E. バッサム著，横山伸也，澤山茂樹，石田祐三郎監訳：エネルギー作物の事典，恒星社厚生閣（2004）
6）伊藤公紀：シリーズ地球と人間の環境を考える1，地球温暖化，日本評論社（2003）
7）奥　彬：シリーズ地球と人間の環境を考える10，バイオマス，日本評論社（2005）
8）昭文社編：なるほど知図帳世界，昭文社（2007）
9）山根一眞：温暖化クライシス，小学館（2006）
10）FAO 2006. FAO Statistics 2006
11）UNDP 2007. HDR Report 2006. United Nations Development Programme

Chapter 8
参考文献
1）中村繁夫：レアメタル資源争奪戦―ハイテク日本の生命線を守れ―，日刊工業新聞社（2007）
2）中村繁夫：レアメタル・パニック―「石油ショック」を超える日本の危機―，光文社（2007）

Chapter 9
参考情報
1）http://www.cbd.int/ (2008)
2）E. O. Wilson : The Diversity of Life, Belknap Press (1992)
3）日高敏隆，総合地球環境学研究所編：子供たちに語るこれからの地球，講談社（2006）

Chapter 10
参考情報
1）http://www.env.go.jp/nature/intro/ (2008)
2）http://www.iucn.org/ (2008)
3）N. Myers : The extinction spasm impending: Synergisms at work. Conservation Biology 1(1), pp.14 〜 21 (1987)
4）http://www.actionbioscience.org/newfrontiers/eldredge2.html (2008)
5）National Research Council USA : Predicting Invasion of Nonindigenous Plants and Pests. National Academy Press (2002)
6）武内和彦，鷲谷いづみ，恒川篤史編：里山の環境学，東京大学出版会（2001）
7）佐藤洋一郎：里と森の危機，朝日新聞社（2005）

Chapter 11
参考情報
1) G. ブレ，N. トルジュマン，L. サン＝マルク著，永田千奈緒訳：地球のやさしいひとになる本，晶文社（2004）
2) アル・ゴア：不都合な真実，ランダムハウス講談社（2007）
3) 田中章義編著：地球では1秒間にサッカー場一面分の緑が消えている，マガジンハウス（2004）
4) 東京商工会議所編：環境社会検定試験eco検定公式テキスト，日本能率協会マネジメントセンター（2006）
5) 林　俊郎：シリーズ地球と人間の環境を考える7，水と健康，日本評論社（2004）
6) Robin Clarke, Jannet King 著，沖　大幹監訳，沖　明訳：水の世界地図，丸善（2006）
7) 国際調査ジャーナリスト協会著，佐久間智子訳：世界の＜水＞が支配される，作品社（2004）
8) 日高敏隆，総合地球環境学研究所編：子供たちに語るこれからの地球，講談社（2006）
9) クリストファー・フレイヴィン編著：地球白書2006-2007，ワールドウォッチジャパン（2005）

Chapter 12
参考情報
1) 岡本裕一朗：異議あり! 生命・環境倫理学，ナカニシヤ出版（2002）
2) 加藤尚武：丸善ライブラリー，応用倫理学のすすめ，丸善（1994）
3) 加藤尚武：丸善ライブラリー，新・環境倫理学のすすめ，丸善（2005）
4) 加藤尚武編：有斐閣アルマ，環境と倫理―自然と人間の共生を求めて―，有斐閣（新版2005（初版は1998））
5) 田上孝一：実践の環境倫理学，時潮社（2006）
6) 谷本光男：環境倫理のラディカリズム，世界思想社（2003）
7) 徳永哲也：はじめて学ぶ生命・環境倫理，ナカニシヤ出版（2003）
8) Jr. Charles E. Harris ほか著，(社)日本技術士会訳編：第2版 科学技術者の倫理，丸善（2002）
9) 加藤尚武・立花　隆 監，山内廣隆 著：現代社会の倫理を考える11，環境の倫理学，丸善（2003）
10) 越智　貢ほか編：岩波 応用倫理学講義2，環境，岩波書店（2004）
11) 加藤尚武・立花　隆 監，村田純一：現代社会の倫理を考える13，技術の倫理学，丸善（2006）
12) 安本教伝編：食の倫理を問う―からだと環境の調和，昭和堂（2000）

英文専門雑誌サイト

Environmental Ethics: http://www.cep.unt.edu/enethics.html (2008)

The International Journal of Environmental, Cultural, Economic and Social Sustainability : http://ijs.cgpublisher.com/ (2008)

International Journal of Sustainable Development: http://www.environmental-expert.com/magazine/inderscience/ijsd/ (2008)

The International Journal of Sustainable Development and World Ecology: http://www.ingentaconnect.com/content/sapi/ijsd (2008)

International Journal of Sustainability in Higher Education: http://www.ingentaconnect.com/content/mcb/249 (2008)

International Society for Environmental Ethics Newsletter (ISEE): http://www.cep.unt.edu/ISEE/index.htm (2008)

Chapter 13
参考情報

1) 今泉みね子：ドイツを変えた10人の環境パイオニア，白水社（1997）
2) デヴィッド・スズキ，ホリー・ドレッセル：グッド・ニュース，ナチュラルスピリット（2006）
3) 深井慈子：持続可能な世界論，ナカニシヤ出版（2005）

Answers to Exercises

Chapter 1
Exercises 1
1. plenty of, 2. pay, 3. purchase, 4. beverage, 5. vend, 6. ordinary, 7. afford,
8. diet, 9. preference, 10. supplementary, 11. spoil, 12. exhaust,
13. unintentional, 14. sufficient, 15. efficient, 16. synthetic, 17. import,
18. agrochemical, 19. maintain, 20. overdose, 21. pesticide, 22. fungicide,
23. pollute, 24. consequence, 25. suffer from, 26. supply, 27. demand,
28. domestic, 29. grain, 30. pulse, 31. handicapped, 32. nation, 33. survival,
34. aliment, 35. transport, 36. end-retail, 37. price, 38. hygiene,
39. accreditation, 40. folk, 41. neglect, 42. dedicate, 43. debate,
44. encourage, 45. competitiveness, 46. integrate, 47. initiation, 48. examine,
49. policy, 50. Trade-off, 51. cumbersome, 52. inconvenient, 53. negotiation,
54. sector, 55. stakeholder, 56. precinct, 57. obesity, 58. metabolic syndrome,
59. revival, 60. rejuvenation, 61. promotion, 62. landscape, 63. conservation,
64. take into account, 65. rural, 66. income, 67. labor, 68. discourage,
69. contemplate on, 70. invest, 71. be proud of

Exercises 2
(1) F, (2) T, (3) F, (4) T, (5) T

Exercises 3
(1) You can purchase plenty of fresh vegetables in the morning market.
(2) We waste enormous amount of energy to supply vegetables and fruit at off-season, and consequently our environment is polluted.
(3) Japanese folks shall debate frequently the international competitiveness of own agriculture.
(4) Let's promote the reduction of garbage.
(5) Every Japanese should complete on the food security.

Exercises 4
(1) Yes, but it is a complex consideration of national policy, local agriculture and rural development, trade balance, consumers recognition and in-

ternational food supply conditions etc.

(2) Yes, but again this requires a lot of political, ethical, legal, economic, social and technological consideration.

(3) Individual preference and thought shall be respected at first. Availability is also another factor. Based on possibility of alternatives, if there are options within Japan including the value, such an import shall be discouraged to buy.

(4) Both are important equally but for the future of our nation, it will be more important the strong diplomatic and negotiation in business as well as our feedback as consumers.

(5) Think about mottainai, "もったいない".

Chapter 2
Exercises 1
1. debate, 2. trans-ocean, 3. recognize, 4. dealer, 5. profit, 6. assure,
7. accredit, 8. standardize, 9. guidance, 10. eel, 11. clam, 12. mackerel,
13. salmon, 14. crab, 15. octopus, 16. tuna, 17. deliver, 18. pandemic,
19. zoonosis, 20. divert, 21. fragile, 22. alternative, 23. import, 24. export,
25. trade, 26. aliment, 27. exhaust, 28. emission, 29. sacrifice, 30. trade off,
31. Out-beat, 32. virtual, 33. indication, 34. consume, 35. consumption,
36. inherent, 37. headache, 38. commodity, 39. entity, 40. essential

Exercises 2
(1) F, (2) F, (3) F, (4) T, (5) T

Exercises 3
(1) Japanese does not only import their foods from all over the world, but also cosume energy and water for the imports at the same time.

(2) Japanese buy a lot of chicken from Brazil as they concern the pandemic breakout of avian influenza to humans.

(3) As far as we depend a lot of food overseas, our food security is fragile.

(4) There are drought and shortage of water-supply even USA which exports food.

(5) We sacrifice our environments by exhausting carbon dioxide in order to produce and transport the aliments.

Exercises 4
(1) It is an overall balance of demand and supply, one sense.
(2) Promotion and patience to use only locally available catches and harvest, or being a vegetarian?
(3) As stated in the text and also go to other chapters.
(4) No. It is totally depending on the trade-off with other factors.
(5) Debates also should be encouraged whether domestic products are superior to imports in term of such environmental concerns, cost, political aspects, trade and social issues.

Chapter 3
Exercises 1
1. relevant, 2. nutritional, 3. value, 4. intake, 5. aliment, 6. supplement,
7. quote, 8. promotion, 9. obesity, 10. diet, 11. disseminate, 12. register,
13. dietitian, 14. implement, 15. care, 16. Tailor-made, 17. expose,
18. consumption, 19. credible, 20. fraud, 21. misinformation,
22. Nutritional pyramid, 23. preference, 24. fiber, 25. mineral, 26. excess,
27. nurture, 28. foundation, 29. pulse, 30. contribution, 31. replace,
32. complementation, 33. physiological, 34. disorder, 35. prevention,
36. affluent, 37. extensive, 38. intervention, 39. behavioral, 40. extension,
41. assessment, 42. evaluation, 43. appropriate, 44. investigation, 45. revise,
46. routine, 47. enhance, 48. surveillance

Exercises 2
(1) F, (2) F, (3) F, (4) T, (5) T

Exercises 3
(1) An appropriate diet helps maintaining a good health.
(2) Obesity is not only attributed to diets and exercises, but also to genetic elements.
(3) Carbohydrates are the foundation for the nutritional pyramid.
(4) Diseases can be prevented by improvement of behavioral activities.
(5) Those supplementary materials cannot replace food and heavy dependency could cause physiological disorders and diseases in a long term.

Answers to Exercises   147

Exercise 4
(1) See the website such as http://www.nih.go.jp/eiken/ and http://fooddb.jp/
(2) List up what you eat daily and consult with dietitian or physician.
(3) It is totally on individual basis and ask for recommendations to specialists.
(4) As in the chapter, food diet is one to consider and other elements to be referred such to http://www.nih.go.jp/eiken/
(5) Yes, indeed, Japanese communities get the support by the educational programs such as Shokuiku.

Chapter 4
Exercises 1
1. Organic, 2. caterpillar, 3. bite, 4. veggie, 5. pesticide, 6. astonish,
7. imaginary, 8. parasitic, 9. herbivore, 10. harmless, 11. nutritious,
12. debate, 13. carcinogenic, 14. bug, 15. horticulture, 16. infestation,
17. pathogen, 18. unedible, 19. fungicide, 20. herbicide, 21. deleterious,
22. regulate, 23. residue, 24. aliment, 25. locality, 26. risk, 27. psychological,
28. tedious, 29. standardization, 30. harmonize, 31. endeavor,
32. examination, 33. inspection, 34. assurance, 35. imaginary fear,
36. Genetic-engineered, 37. compulsory, 38. commodity, 39. broadcast,
40. sensational, 41. falsify, 42. diarrhea, 43. pulse, 44. audience, 45. reliable,
46. trustable

Exercises 2
(1) F, (2) F, (3) T, (4) T, (5) T

Exercises 3
(1) Imaginary fears and concerns come up on modern technologies such as food additives and biotechnology-derived food.
(2) Consumers were astonished that some food processors falsified the content information.
(3) The veggies bitten by bugs are not necessarily unedible.
(4) We ought to understand where we import food stuffs.
(5) The inadequate broadcasting confused the audience.

Exercises 4
(1) You judge by yourself at your risk?
(2) Public organizations furnish relevant information for your help.
(3) It is again your judgment.
(4) It is endless but always sharing ideas is important to increase occasions on dialogue exchange and expose yourself to other options.
(5) You are the final decision-maker.

Chapter 5
Exercises 1
1. conflict, 2. existence, 3. betterment, 4. continuity, 5. essential, 6. creature, 7. brine, 8. saline, 9. halophytic, 10. microorganism, 11. tolerant, 12. quench, 13. flush, 14. realize, 15. allocation, 16. practice, 17. proportion, 18. purify, 19. reusable, 20. reservoir, 21. contribution, 22. implementation, 23. irrigation, 24. deplete, 25. salinization, 26. fertility, 27. prominent, 28. dependency, 29. concentration, 30. capillary motion, 31. suck up, 32. well, 33. landscape, 34. deposit, 35. destruct, 36. domestic, 37. aquatic, 38. recovery, 39. shortage, 40. rain fall, 41. pollution, 42. concentrate, 43. upstream, 44. downstream, 45. sanitation, 46. hygiene, 47. incorporate, 48. reduction, 49. mortality, 50. epidemic, 51. fundamental, 52. coexistence

Exercises 2
(1) F, (2) T, (3) F, (4) T, (5) T

Exercises 3
(1) Stockholm Environmental Institute integrates and maps the information globally on conflict and problematic sites associated with water.
(2) Sufficient supply of clean water is essential for reduction of the mortality on babies and infants.
(3) It can be deemed that salinization of soil is attributed to the concentration of saline by capillary motion with sucking a lot of water up from deep wells.
(4) Agriculture practices require enormous amount of water, and exert pressures on demands from urban life.
(5) UN Millennium Development Goals raise eight targets to attain, and

water is the important component for the proposed achievements.

Exercises 4

(1) Reduction of virtual water is one and other political efforts should be considered on the international water issues.

(2) This is the same as other natural resources: reduction of use, avoiding pollution and reuses such by bathing water for washing and flushing toilets would be easy to start.

(3) It is depending on the given situations. More importance is on how we can avoid further causes and incidences.

(4) The problems are the consequences of a bad combination of policy and unbalance between productivity and sustainability.

(5) Yes, but further endeavors are essential in science and technology to make efficiency as making fresh water from brine needs enormous energy and cause emission problems. Biotechnology can also help but social acceptance and comprehensive regulatory systems are the bottlenecks for the spread of the genetically-engineered trees and crops which can grow under harsh saline conditions, although they are ready for planting to reduce fresh water uses.

Chapter 6
Exercises 1

1. waste, 2. garbage, 3. rubbish, 4. stuff, 5. urban, 6. rural, 7. dump,
8. designate, 9. sanitary, 10. compost, 11. transform, 12. fossil, 13. edible,
14. expose, 15. hunger, 16. starvation, 17. affluent, 18. sophistication,
19. satisfy with, 20. precious, 21. ruin, 22. transportation, 23. handle,
24. process, 25. storage, 26. promotion, 27. dwindle, 28. recycle, 29. reuse,
30. reduce, 31. disinfection, 32. enforce, 33. abandonment, 34. manufacturer,
35. retailer, 36. end-user, 37. supplier, 38. purchase, 39. furnish, 40. trash,
41. bury, 42. pile up, 43. revive, 44. packaging, 45. wrapping,
46. shrimp, prawn, 47. coastal, 48. antibiotics, 49. pitfall, 50. melt

Exercises 2

(1) F, (2) T, (3) T, (4) T, (5) F

Exercises 3

(1) It is the idiotic act that dumping wastes at your precinct ridges is not only illegal but also inconsiderate on environmental ethics and social view points.

(2) Food shall be purchased and consumed with a plan.

(3) There are always many people suffered by starvation worldwide.

(4) It is prohibited to abandon the cars and domestis electric items by yourself without proper procedures.

(5) It is possible to decrease domestic garbage and trash by individual consideration.

Exercises 4

(1) Environmentally speaking, it is yes. But with economics view point, there shall be a replacement of industry to complement the income opportunities or be a sustainable way of the production.

(2) A stated in the text and as you think.

(3) Yes, and do not be discouraged with the unsophisticated statements or conducts by blunt people.

(4) Governance is cardinal while individual efforts are essential: the amalgamated package of Policy, regulation, systems, social activities such as public enlightenment and science and technology for alleviating issues.

(5) Consumers wisdom is respected.

Chapter 7

Exercises 1

1. skyrocket, 2. transportation, 3. commodity, 4. movement, 5. motor vehicle,
6. emission, 7. occasion, 8. life-line, 9. firewood, 10. over-exploitation,
11. cardinal, 12. deforestation, 13. divert, 14. fossil, 15. consume, 16. scream,
17. petroleum, 18. nuclear power, 19. atomic energy, 20. irradiation,
21. release, 22. skeptical, 23. abuse, 24. mismanagement, 25. catastrophe,
26. nuclear fusion, 27. achieve, 28. solar, 29. sea tide, 30. Stone Age,
31. reproducible, 32. promote, 33. combustible, 34. boost, 35. starch,
36. fatty acid, 37. lipid, 38. utilization, 39. forecast, 40. indulge in,
41. tremendously, 42. diversion, 43. tortilla, 44. flour, 45. delicacy, 46. retail,

47. riot, 48. decomposition, 49. cellulose, 50. compound, 51. lignin, 52. architecture, 53. fermentation, 54. Genetic engineering, 55. elaboration, 56. investment

Exercises 2

(1) F, (2) F, (3) T, (4) T, (5) T

Exercises 3

(1) His research achievement was abused by terrorists.

(2) If we can use no energy, our life would go back to the way at the Stone Age.

(3) Atomic energy can generate a lot of electricity but management is cardinal for safety assurance.

(4) Plant metabolic functions can be elaborated by genetic engineering, and production of fatty acids can be facilitated.

(5) Exploitation of options on the energy sources is the cardinal investment for human future.

Exercises 4

(1) It is totally depending on where and how the materials are produced and processed. Also there is yet a lot of challenge needed for effective uses of cellulose and lignin on bio-ethanol and for increased efficiency on bio-diesels.

(2) Pass the buck to the president? It is very complex to make such a balance. But it is easy to say that food sources should not be used for alternative purposes, and options on energy sources should be examined.

(3) It is good to engage for future contributions to human societies and the planet, but one should realize that it is tedious toward a lot of trials and errors like any other field of science and technology.

(4) No, it is yet an important sources for large demands.

(5) No, it totally depending on how the integrated system is well made for sustainable uses.

Chapter 8
Exercises 1

1. rare metal, 2. element, 3. entity, 4. product, 5. cellular phone, 6. property,

152   Answers to Exercises

7. furnish, 8. extraction, 9. purification, 10. concentration, 11. scarce, 12. parenthesis, 13. atomic symbol, 14. atomic number, 15. rare earth, 16. mine, 17. oxidize, 18. copper, 19. antimony, 20. zinc, 21. tremendously, 22. heat tolerance, 23. resistance to corruption, 24. magnetic, 25. fluorescent, 26. compound, 27. alloy, 28. ignition, 29. semiconductor, 30. ceramic, 31. superconductive, 32. illuminate, 33. platinum, 34. geographical, 35. instability, 36. fluctuation, 37. nickel, 38. aluminum, 39. tungsten, 40. manufacture, 41. deleterious, 42. resources security, 43. diplomatic negotiation, 44. tedious, 45. acquire

Exercises 2

(1) T, (2) F, (3) T, (4) T, (5) T

Exercises 3

(1) The use of cellular phones is sky-rocketing, and there is an increasing concern on the supply of the rare metals as the materials for the handy phones.

(2) Rare earth elements have similar property.

(3) Rare metals are employed also in highly processed alloys and medical devices.

(4) Barium and gadolinium are used as contrast reagents, for X-ray and for MRI, respectively.

(5) There are more rigid regulatory systems on the exploitation of the metal resources by foreign investment and it takes more complicated in diplomatic negotiation and civilian business talks.

Exercises 4

(1) Rare metals are a group of elements consisting of 31 different entities. They are industrially important materials for high quality products such as cellular phones.

(2) Seventeen entities are called rare earth such as scandium, (Sc, 21), yttrium (Y, 39) neodymium (Nd, 60) and terbium (Tb, 65). The latter two belong to Lanthanide series.

(3) Compounds and alloys made from them are used for ignition alloys, lenses, fluorescent items, laser sources, permanent magnet and base materials for computer semiconductor. Ceramics containing the rare earth can have high temperature superconductive capacity and are employed at different advanced industrial products.

(4) They have common properties such as heat tolerance, resistance to corruption, magnetic capacity and fluorescent activity. These materials also help smaller-sizing, reducing weight and higher performance of industrial products as well as energy-use efficiency. Japanese high technology industry had success with the rare metals, and there is no further growth without them.

(5) Yes, platinum is a good example. Automobile industry consume the sixty percent of world demands on platinum. Each car needs three gram of platinum. The up-coming energy cell operated car requires about 80 gram per vehicle.

Chapter 9
Exercises 1
1. food chain, 2. biodiversity, 3. mixed breed (hybrid), 4. piebald, 5. fur,
6. silly (stupid), 7. urban, 8. rural, 9. precinct (neighborhood), 10. territory,
11. exploit, 12. nectar, 13. predator, 14. reptile, 15. lizard, 16. taxonomic,
17. evergreen, 18. deciduous, 19. needle-leaved, 20. broad-leaved, 21. bush,
22. shrub (bush), 23. autumn tint, 24. branch, 25. excrement, 26. eelworm,
27. mole, 28. cohort, 29. genetics, 30. inheritance, 31. endangered,
32. paddy, 33. ecosystem, 34. arrowhead, 35. japanese barn millet,
36. mud snail, 37. loach, 38. heron, 39. drastic, 40. marsh, 41. landscape,
42. erosion, 43. creature, 44. aquatic, 45. amphibian, 46. crawfish,
47. salamander, 48. dragonfly, 49. nymph, 50. population, 51. abundant,
52. proliferate, 53. dwindle, 54. scarce, 55. rare, 56. domesticated, 57. cicada,
58. waste, 59. perpetuate, 60. recurrent, 61. reincarnation, 62. ethical,
63. religious

Exercises 2
(1) T, (2) T, (3) F, (4) F, (5) F

Exercises 3
(1) Poo observed a lizard at a branch of a evergreen tree.

(2) Autumn tints can be seen about from March to May in such a country

like New Zealand at Southern hemisphere.
(3) The body color of medaka follows Mendelian inheritance.
(4) Biodiversity has been dwindling due to the global warming and the environmental destructions.
(5) A creature at the peak of a food chain can perpetuate its species.

Exercises 4
(1) Evergreens maintain their leaves all the seasons such as needle-leaved conifers and pines. Deciduous ones such as apples and birches, defoliate their leaves each fall and rejuvenate the leaves in each spring.
(2) No, they are in the different categories. Salamanders are amphibians and lizards are reptiles.
(3) No, a lot of efforts and compromises should be made to get along with the nature by mankinds.
(4) It is up to you by your creed, religion and value on your life.
(5) Yes, a consideration should be made on the environmental ethics and bioethics views when we contemplate on the biodiversity conservation.

Chapter 10
Exercises 1
1. Invasive(ness), 2. crisis, 3. migration, 4. speciation, 5. evolutionary,
6. propagule, 7. occurrence, 8. phenomenon(na), 9. disturbance,
10. extinction, 11. spasm, 12. catastrophe, 13. cataclysm, 14. speculate,
15. meteorite, 16. strike, 17. evolve, 18. climate, 19. overexploitation,
20. pollution, 21. irreversible, 22. alleviation, 23. enforce, 24. enormous,
25. mobilize, 26. intentionally, 27. culinary, 28. plantain, 29. dwindle,
30. drastic, 31. scenery, 32. evoke, 33. nursery rhyme, 34. dandelion,
35. hybridization, 36. endangered, 37. extinguish, 38. extinct, 39. threaten,
40. consequence, 41. concern, 42. deliberate, 43. catfish, 44. ruin,
45. herbaceous, 46. substance, 47. allelopathic, 48. fecundity, 49. pollen,
50. insect-borne, 51. reproduction, 52. propagation, 53. vegetation,
54. aquarium, 55. explosively, 56. adaptability, 57. degradation, 58. survive,
59. protozoa, 60. infectious, 61. parasitic, 62. quarantine, 63. pathogen(ic),

64. abandonment, 65. regulate, 66. prejudice, 67. indigenous,
68. alien/exotic, 69. risk, 70. benefit

Exercises 2

(1) F, (2) F, (3) T, (4) T, (5) T

Exercises 3

(1) Canada golden-rod had invaded the corresponding ecosystems in Japan, China and European countries, and the vegetation have been threatened.

(2) Conservation of biodiversity can be enforced by primary education and public awareness.

(3) Importation of alien species is regulated by several laws.

(4) Consequences by human activity to biodiversity, show up in a short span not at a distant future.

(5) It is impossible to regenerate an extinct species, and it is irreversible to recover a destroyed ecosystem.

Exercises 4

(1) See the text.

(2) Many paleontologists and archaeologists speculate that a meteorite fallen to the earth might have made huge environmental changes in the past, and ruined the dinosaurs.

(3) We should consider our daily life style such as use of energy sources which highly associate with global warming.

(4) Yes, crop exchange made innovation and development of human life.

(5) No. Only with scientific risk assessment, it could be judged.

Chapter 11

Exercises 1

1. degradation, 2. consist of, 3. inanimate, 4. animate, 5. atmosphere,
6. terrain, 7. organism, 8. moss, 9. fern, 10. fungus, 11. bacterium,
12. mammalian, 13. creation, 14. unpredictable, 15. natural disaster,
16. deteriorate, 17. agitate, 18. worsen, 19. landscape, 20. landslide,
21. contaminate, 22. pollute, 23. emission, 24. stuff, 25. boost,
26. indiscriminately, 27. disturabance, 28. destruction, 29. diversion,

30. process, 31. spare, 32. neglect, 33. earthquake, 34. volcano, 35. eruption, 36. erosion, 37. salinization, 38. deforestation, 39. desertification, 40. melt, 41. glacier, 42. pole, 43. decline, 44. arable, 45. barren, 46. sink, 47. coral reef, 48. coast(al), 49. sadness, 50. unbalance, 51. forecast, 52. pessimistic

Exercises 2

(1) F, (2) F, (3) F, (4) F, (5) T

Exercises 3

(1) Inanimate and animate beings interact and they form up a variety of ecosystems leading to creation of biological diversity.

(2) These are serious damages to conservation of landscape and biological diversity.

(3) Invasive species like tilapia does not have only a risk to extinct individual domestic aquatic species, but also it could destroy totally the ecosystem such by water lilies.

(4) It is not feasible to predict the natural incidents, however, it shall be sufficient precautionary approach and preparation for them.

(5) We shall think about saving resources and protecting our environment by our daily activities.

Exercises 4

(1) No, those species which are invasive at one ecosystem, does not necessary the same at different conditions. Also think about ethical components and the wisdom of human beings, the could be options to cope with the presence of the invasive species.

(2) We should not be optimistic nor pessimistic. Think before making actions to the best efficiency of uses of convenient items which lead to the emission.

(3) Promotion of dialogue exchange is one, and also any of participatory collaboration is another to understand and help each other.

(4) We always should think globally and act locally.

(5) Please debate more and show your actions.

Chapter 12
Exercises 1
1. ethics, 2. discipline, 3. emerge, 4. comprehensive, 5. philosophy,
6. sociology, 7. theology, 8. economics, 9. ecology, 10. geography,
11. aesthetics, 12. branch, 13. investigation, 14. sustainability, 15. fairness,
16. equity, 17. stakeholder, 18. argument, 19. forum, 20. resident,
21. accommodate, 22. ecologist, 23. genome, 24. biotechnology, 25. influence,
26. conservative, 27. chemical, 28. property, 29. component,
30. multidisciplinary, 31. approach, 32. duty, 33. generation, 34. forestry,
35. respectful, 36. mysterious, 37. resources, 38. game, 39. carrier,
40. realize, 41. incline, 42. deteriorate, 43. state, 44. advocate,
45. preservation, 46. wildlife, 47. wilderness, 48. raise, 49. serious,
50. synthetic, 51. pesticide, 52. extinction, 53. pioneer, 54. ecocentrism,
55. biocentrism, 56. anthropocentrism, 57. priority, 58. point,
59. cohabitation, 60. limitation, 61. implication, 62. benefit, 63. publication,
64. reality, 65. wealth, 66. poverty, 67. humanitarian, 68. degradable,
69. selfishness, 70. abuse, 71. bioethics, 72. exterminate, 73. emission,
74. conceptual, 75. dedication

Exercises 2
(1) T, (2) T, (3) F, (4) T, (5) F

Exercises 3
(1) We must realize that new environmental problems emerge consecutively.
(2) Leopold respected wildlife and wilderness and he advocated environmental ethics.
(3) It should be raised that appropriate uses of pesticides do not always cause serious environmental hazards.
(4) She inclined to the environmental preservation with the view point of ecocentrism.
(5) It should be pointed out that the priority setting is essential to deal with the environmental issues.

Exercises 4
(1) we hope so, but consequence can be realized after a far future.

(2) Yes, but it is very difficult to consider all thought equally.
(3) You can debate the implication.
(4) Yes or No, but it cannot be a simple answer but the ethical consideration is cardinal.
(5) Go over this chapter.

Chapter 13
Exercises 1
1. waste, 2. modernized, 3. turn off, 4. switch on, 5. consume, 6. save, 7. spoil, 8. thermal power generation, 9. plant, 10. nuclear, 11. auxiliary, 12. co-generation, 13. divert, 14. skyrocketing, 15. reduce, 16. rotate, 17. house-keeping, 18. compost, 19. replacement, 20. accumulate, 21. legal, 22. retailer, 23. burn, 24. deal, 25. dispose, 26. disposal, 27. garbage, 28. recyclable, 29. recycle, 30. reproduction, 31. reuse, 32. manufacture, 33. encourage, 34. restrict, 35. compliance, 36. contribute, 37. traffic jam, 38. productivity, 39. materialize, 40. exploit, 41. harvest, 42. elaboration, 43. implementation

Exercises 2
(1) F, (2) T, (3) F, (4) T, (5) F

Exercises 3
(1) It is not unwise to abuse the abundant natural resources.
(2) Food should not be spoiled, but it shall be imported and supplied with an appropriate plan.
(3) Nuclear power can generate electricity at a reasonable efficiency.
(4) Sustainable use of natural resources heavily depend on recycle, reuse and reduction of excess use.
(5) It is highly encouraged to make the energy saving at your ordinary life.

Exercises 4
(1) Yes, but there is prerequisite in infrastructure and associated system together with a local regulatory and social coordination.
(2) You can take examples and give your own direction from this book.
(3) Compliance is vital as a corporation and as individuals.

(4) Individualism may be an obstacle, but participatory arrangements and some regulatory compulsion shall change drastically for spreading the activity.

(5) As you wish.

Index

【 A 】

abandonment	48
abundant	77
abuse	56
abusing	107
accommodate	105
accredit	11
accreditation	4
accumulate	116
accuracy	23
achieve	56
acid	57
acquired	70
adaptability	88
aesthetics	104
affluent	21, 41, 46
afford	1
agenda	122
agitate	94
agrochemical	2
alien	88, 91, 92, 93
aliment	3, 11, 13, 19, 29
allelopathic	87
allergic	28
alleviate	5, 29
alleviation	83, 117
allocation	38
alloy	66
alternative	12
aluminum	69
ambitious	123
amphibian	77
animate	94
anthropocentrism	106
antibiotic	49
antimony	65
approach	105
aquarium	87
aquatic	40, 77
arable	40, 96
Aral Sea	38
architecture	58
argument	105
arrowhead	76
artificial	1
assessment	21
assurance	29, 30
assure	11
astonish	28
atmosphere	94
atomic energy	55
atomic number	65
atomic symbol	65
audience	31
autumn tint	75
auxiliary	114
avian influenza	12

【 B 】

bacteria	94
barium	68
barren	40, 96
behavioral	21
benefit	89, 93, 106
betterment	36
beverage	1
bias	31
biocentrism	106
biodiversity	75
bioethics	107
biotechnology	105
bite	28
bitterling	85
black bass	85
bluegill	85
boost	57
boosting	95
branch	75
brassica	84
brine	36
broadcast	31
broad-leaved	75
BSE	12
bug	28
burnt	116
bury	49
bush	75

【 C 】

canada golden-rod	87
canola oil	11
capacity	123
capillary	40
carcinogenic	28
cardinal	55
care	20
carp herpes virus	85
carrier	105

Index   **161**

cataclysm	83	consumption	14, 20	demonstrate	123
catastrophy	56, 83	contaminate	94	dependency	38
caterpillar	28	contamination	28	deplete	38
catfish	85	contemplate on	6	deposite	40
cellular phone	65	continuity	36	depress	5
cellulose	58	contribution		desertification	96
ceramic	66		21, 38, 116, 122	design	19
chemical	105	cool-biz	99	designate	46
chromium	69	copper	65	destruct	40
cicada	78	coral reef	96	destruction	95
clam	12	crab	12	deteriorate	105
climate	83	crawfish	77	deteriorating	94
coastal	49, 96	creation	94	diarrhea	31
coexistence	41, 123	creation	123	dictate	107
co-generation	114	creature	37, 77	diet	1, 19
cohabitation	106	credible	20	dietary	19
cohort	21, 76	crisis	83	dietitian	19
collaboration	122	critical	15	diplomatic negotiation	69
combustible	57	CSR	116	discipline	104
commodity	31, 55	culinary	84	discouragement	6
common reed	77	cumbersome	5	disinfection	46
competitiveness	4	cyclone	96	disposal	116
complementation	21	**[D]**		dissemination	19
compliance	116			disturbance	83, 95
component	5, 105	damage	38	diversion	60, 95
compost	46, 116	dandelion	85, 86	divert	12, 55, 114
compound	58, 66	dealer	11	domestic	2, 40
comprehensive	104	dealt	116	domesticate	78
compulsory	30	debate	4, 11, 28	downstream	41
concentrate	41	deciduous	75	dragonfly	77
concentration	38	decline	96	drastic	77, 85
conceptual	108	decomposition	58	drastically	38
concern	85	dedication	108	DRIs	23
conflict	36	deforestation	55, 96	dump	46, 48
consensus	31	degradation	88, 94, 107	duty	105
consequence	2, 85	deleterious	29, 69	dwindle	38, 50
conservation	6, 15	deliberate	85	dwindling	46, 77, 84
conservative	105	delicacy	58	**[E]**	
consist	94	deliver	12		
consume	14, 55, 114	demand	2	earthquake	96

ecocentrism	106	exceed	117	forestry	105
ecologist	105	excess	21	forum	105
ecosystem	76	excrement	76	fossil	46, 57
edible	46	exhaust	1, 12	foundation	20
eel	12	exhaustion	123	fragile	12
eelworm	76	existence	36	fraud	20
efficient	1	exotic	88, 89	functional	19
elaborat	58	exploite	117	fundamental	41
elaboration	117	exploiting	75	fungi	94
element	65	explosively	87	fungicide	2, 29
emerging	104	export	12	fur	75
emission	13, 55, 95, 108	expose	20, 46	furnish	49, 65
encourage	4, 116	extensive	21	**【 G 】**	
encouragement	4	exterminate	107		
endanger	76, 85, 87	extinct	87, 92	gadolinium	68
endangered	86	extinction	106	gallium	66
endeavor	30	extinction spasm	83	game	105
end-retail	4	extinguish	85	garbage	49, 116
end-user	48	extraction	65	generation	105
enforce	48, 123	**【 F 】**		genetic	23, 76
enforced	83			genetic engineering	58
enhance	23	face	41	genetic-engineered	30
enrich	104	fairness	105	genome	105
enormous	15, 84	falsify	31	geographical	68
entity	15, 65	fatty	57	geography	104
epidemic	41	fecundity	87	glacier	37, 96
epilogue	122	fermentation	58	grain	3
equity	105	fern	94	guidance	11
erosion	77, 96	fertility	38	**【 H 】**	
eruption	96	fiber	21		
essential	15, 36	firewood	55	HACCP	11
ethical	78, 104	flour	58	halophytic	37
ethics	105	fluctuation	69	handicappe	3
Euphrates River	41	fluorescent	66	handle	46
evaluation	21	flush	37	harmless	28
evergreen	75	folk	4	harmonize	30
evoke	85	food chain	75	harvest	117
evolutionary	83	food faddism	31	headache	15
evolving	83	foot-and-mouth disease	29	heat tolerance	66
examination	30	forecast	57, 98	heliotrope	115

Index　163

herbaceous	87	initiation	4	legal	116
herbicide	29	insect-borne	87	legal instrument	104
herbivore	28	insecticide	28	legend	104
heritage	123	inspection	30	life-line	55
heron	76	instability	68	lignin	58
horticultural	28	intake	19	limitation	106
house-keeping	116	integrate	4	lipid	57
humanitarian	106	integrative	123	lizard	75
hunger	46	intentionally	84	loach	76
hurricane	96	interpretation	31	locality	29
hybridization	85	intervention	21	logic	104
hygiene	4, 40	invasive　83, 86, 87, 88, 89, 91, 93		**【 M 】**	
【 I 】		invasiveness	87, 88, 89	mackerel	12
IAEA	56	invest	6	magnetic	66
ignition	66	investigate	21	maintain	1
illness	19	investigation	23	mammalian	94
illuminating	66	investment	58	manufacture	69
imaginary	28	IPCC	57, 108	manufacturer	48
imaginary fear	30	irradiation	55	manufacturing	116
implement	19, 21	irreversible	83, 85, 92	marsh	77
implementation	29, 38, 117	irrigation	38	materializing	117
import	1, 12	IUCN	87	melt	50
inanimate	94	ISO	116	melting	96
incline	105	**【 J 】**		metabolic syndrome	5
income	6			meta physical	104
inconvenience	5	Japanese crested ibis	76	meteorite	83, 93
incorporate	41	Japanese barn millet	76	microorganism	37
independent	40	JAS	31	migration	83, 91
indication	14	**【 K 】**		mine	65
indigenous	87			mineral	21
indiscriminately	95	Kyoto Protocol	57,108	misinformation	20
indium	68	**【 L 】**		mismanagement	56
indulged in	57			mixed breed	75
infant	40	labor	6	mizuaoi	77
infectious	28, 88	landscape	6, 40, 77, 94	mobilize	84
infestation	28	landslide	94	modernize	114
influence	105	lanthanide series	65	mole	76
inherent	15	LCD	68	molybdenum	69
inheritance	76	LED	66	mortality	41

moss	94	organism	94	population	77	
motion	40	orient	123	poverty	106	
motor vehicle	55	oriental stork	76	prawn	49	
movement	55	overdose	2	precinct	5, 75	
MRI	68	overexploitation	38, 55, 83	precious	46	
mud snail	76	oxidize	65	predator	75	
multidisciplinary	105			preference	1, 20	
mysterious	105	**[P]**		prejudice	89	
		package	49	preservation	106	
[N]		paddy	76	prevent	21	
nation	3	pandemic	12	price	4	
natural disaster	94	parasitic	28, 88	priority	106	
nectar	75	parenthesis	65	problematic	41	
needle-leaved	75	pathogen	28	process	12, 46, 49, 50	
neglect	4, 95	pathogenic	88	processing	95	
negotiation	5	patience	5	producer	31	
neodymium	65	pay	1	product	65	
nickel	69	permanent	37	productivity	117	
nile parch	95	perpetuating	78	profit	11	
no abandonment	88	pessimistic	98	proliferate	77	
nuclear	114	pesticide	2, 28, 106	prominent	38	
nuclear fusion	56	petroleum	57	promote	5, 57	
nuclear power	55	phenomena	83	promotion	19, 46	
nursery rhyme	85	physiological	21	propagation	87	
nurture	21	phytoalexin	28	propagule	83	
nutraceutical	19	piebald	75	property	65, 105	
nutritional	19	pile up	49	proportion	38	
nutritional pyramid	20	pitfall	50	protozoa	88	
nutritious	28	planet	36	psychological	29	
nymph	77	plant	114	publication	107	
		plantain	84	pulse	3, 21, 31	
[O]		plasma TV	68	purchase	1, 49	
obese	21	platinum	66	purification	65	
obesity	5, 19	plenty of	1	purify	38	
occasion	55	point	106			
occurrence	83	poison	28	**[Q]**		
octopus	12	pole	96	quarantine	88	
omodaka	76	pollen	87	quench	37	
ordinary	1	pollute	2, 15, 40, 94	quote	19	
organic	28	pollution	41, 83			

Index

[R]

rain fall	40
raise	106
rare	77
rare earth	65
rare metal	65
R&D	29
reality	106
realize	15, 38, 105
recognize	11
recovery	40
recurrent	78
recycle	46, 50
recycling	116
reduce	49, 50, 114
reducing	114
reduction	41
refrain	5
register	19
regulate	29, 88, 92
regulator	31
reincarnation	78
rejuvenation	6, 123
release	55
relevant	19
reliable	31
relief	28
remediation	122
replace	21
replacement	116
representation	123
reproducible	56
reproduction	87, 116
reptile	75
reservoir	38
resident	105
residential	123
residue	29
resistance to corruption	66
resource	105
resources security	69
respectful	105
restricted	116
restructure	123
retail	58
retailer	48, 116
reusable	38
reuse	46
reusing	116
revise	23
revival	5
revive	49
riot	58
risk	29, 89
rotate	115
routine	23
rubbish	46
ruin	46, 87
rural	6, 46, 75

[S]

sacrify	13
sadness	96
salamander	77
saline	36
salinization	38, 96
salmon	12
sanitary	46
sanitation	40
satisfied with	46
saving	114
scandal	31
scandium	65
scarce	65, 77
scenery	85
scream	55
sea tide	56
selfishness	106
semiconductor	66
sensational	31
serious	106
shokuiku	19
shortage	40
shrimp	49
shrub	75
silly	75
sinking	96
skeptical	56
skyrocketing	55, 114
slash-and-burn	94
solar	56
sophisticated	31
sophistication	46
sparing	95
spasm	91, 92, 93
special dietary food	19
speciation	83, 91
speculate	83
spoil	1
spoiled	114
spread	123
stakeholder	5, 31, 105
standardization	30
standardize	11
starchy	57
starvation	46
state	105
Stone Age	56, 62
storage	46
struck	83
stuff	46, 50, 95
subprogram	122
subsequently	5
subside	4
substance	28, 87
suck	40
suffer from	2
sufficient	1, 40
superconductive	66

supplement	19	transform	46	veggie	28
supplementary	1	trans-ocean	11	vend	1
supplier	31	transport	3	virtual	14
supply	2, 49	transportation	46, 55	vital	28
surveillance	23	trash	49	volcano	96
survival	3	tremendously	58, 66		
survive	88	trial	123	**【 W 】**	
sustainability	41, 105	trustable	31	warm-biz	99
sustainable	6	tuna	12	waste	46, 78
synthetic	1, 106	tsunami	96	wasted	114
		tungsten	69	water chestnut	77
【 T 】		turn off	114	wealth	106
tailor-made	20	Tuvalu	96	well	40
taking into account	5			wilderness	106
task	117	**【 U 】**		wildlife	106
taxonomic	75	unbalance	98	worsen	94
tedious	29, 70	UNCED	108	wrap	49
terbium	65	undernourish	20	WSSD	108
terrain	94	undertake	23	WTO	5
territory	75	unedible	29		
theme	122	unintentional	1	**【 X 】**	
theology	104	unpredictable	94	X-ray	68
thermal power generation	114	unrecyclable	116	**【 Y 】**	
		upstream	41		
threaten	87	urban	46, 75	yttrium	65
tirapia	95	utilization	57	Yunnan Province	41
tolerant	37				
tortilla	58	**【 V 】**		**【 Z 】**	
trade	12	value	19	zinc	65
trade-off	4, 13	vanadium	69	zoonosis	12
traffic jam	116	vegetation	15, 87		

―――― 編著者・著者略歴 ――――

渡邉　和男（わたなべ　かずお）
- 1983 年　神戸大学農学部園芸農学科卒業
- 1985 年　神戸大学大学院修士課程修了
- 1988 年　ウィスコンシン大学大学院博士課程修了（遺伝育種学専攻）
　　　　　Ph. D.（ウィスコンシン大学）
- 1988 年　国際ポテトセンター主任研究員
- 1991 年　コーネル大学助教授
- 1996 年　近畿大学助教授，国際植物遺伝資源研究所（IPGRI）名誉研究員
- 2001 年　筑波大学教授，コーネル大学在外特別教授
- 2004 年　筑波大学大学院教授
　　　　　現在に至る

渡邉　純子（わたなべ　じゅんこ）
- 1986 年　神戸大学教育学部初等教育学科卒業
- 1986 年　兵庫県立須磨東高等学校教諭
- 1995 年　コーネル大学大学院修士課程修了（園芸，植物生理学専攻）
- 1997 年　近畿大学非常勤講師
- 2001 年　筑波大学遺伝子実験センター研究推進員
- 2005 年　農業生物資源研究所非常勤職員
- 2007 年　科学作家
　　　　　現在に至る

英語で学ぶ環境科学
― 食料，資源と環境 ―

Introduction to Environmental Sciences and Practices
― Food, Natural Resources and Environment ―

Ⓒ　Kazuo Watanabe, Junko Watanabe　2008

2008 年 8 月 1 日　初版第 1 刷発行

検印省略	編著者	渡　邉　和　男
	著　者	渡　邉　純　子
	発行者	株式会社　コロナ社
		代表者　牛来辰巳
	印刷所	萩原印刷株式会社

112-0011　東京都文京区千石 4-46-10

発行所　株式会社　コロナ社

CORONA PUBLISHING CO., LTD.

Tokyo　Japan

振替 00140-8-14844・電話(03)3941-3131(代)

ホームページ　http://www.coronasha.co.jp

ISBN 978-4-339-07882-4　（高橋）　（製本：愛千製本所）
Printed in Japan

無断複写・転載を禁ずる

落丁・乱丁本はお取替えいたします

地球環境のための技術としくみシリーズ

(各巻A5判)

コロナ社創立75周年記念出版

■編集委員長　松井三郎
■編集委員　小林正美・松岡　譲・盛岡　通・森澤眞輔

配本順				頁	定価
1. (1回)	今なぜ地球環境なのか	松井三郎編著		230	3360円
	松下和夫・中村正久・髙橋一生・青山俊介・嘉田良平 共著				
2. (6回)	生活水資源の循環技術	森澤眞輔編著		304	4410円
	松井三郎・細井由彦・伊藤禎彦・花木啓祐 荒巻俊也・国包章一・山村尊房 共著				
3. (3回)	地球水資源の管理技術	森澤眞輔編著		292	4200円
	松岡　譲・髙橋　潔・津野　洋・古城方和 楠田哲也・三村信男・池淵周一 共著				
4. (2回)	土壌圏の管理技術	森澤眞輔編著		240	3570円
	米田　稔・平田健正・村上雅博 共著				
5.	資源循環型社会の技術システム	盛岡　通編著			
	河村清史・吉田　登・藤田　壮・花嶋正孝 宮脇健太郎・後藤敏彦・東海明宏 共著				
6. (7回)	エネルギーと環境の技術開発	松岡　譲編著		262	3780円
	森　俊介・槌屋治紀・藤井康正 共著				
7.	大気環境の技術とその展開	松岡　譲編著			
	森口祐一・島田幸司・牧野尚夫・白井裕三・甲斐沼美紀子 共著				
8. (4回)	木造都市の設計技術			282	4200円
	小林正美・竹内典之・髙橋康夫・山岸常人 外山　義・井上由起子・菅野正広・鉾井修一 吉田治典・鈴木祥之・渡邊史夫・高松　伸 共著				
9.	環境調和型交通の技術システム	盛岡　通編著			
	新田保次・鹿島　茂・岩井信夫・中川　大 細川恭史・林　良嗣・花岡伸也・青山吉隆 共著				
10.	都市の環境計画の技術としくみ	盛岡　通編著			
	神吉紀世子・室崎益輝・藤田　壮・鳥谷幸宏 福井弘道・野村康彦・世古一穂 共著				
11. (5回)	地球環境保全の法としくみ	松井三郎編著		330	4620円
	岩間　徹・浅野直人・川勝健志・植田和弘 倉阪秀史・岡島成行・平野　喬 共著				

定価は本体価格+税5%です。
定価は変更されることがありますのでご了承下さい。

図書目録進呈◆

シリーズ　21世紀のエネルギー

(各巻A5判)

■(社)日本エネルギー学会編

	書名	著者	頁	定価
1.	21世紀が危ない ― 環境問題とエネルギー ―	小島　紀徳著	144	1785円
2.	エネルギーと国の役割 ― 地球温暖化時代の税制を考える ―	十市　勉 小川　芳樹 共著 佐川　直人	154	1785円
3.	風と太陽と海 ― さわやかな自然エネルギー ―	牛山　泉他著	158	1995円
4.	物質文明を超えて ― 資源・環境革命の21世紀 ―	佐伯　康治著	168	2100円
5.	Cの科学と技術 ― 炭素材料の不思議 ―	白石　共 京谷・大山　著 谷田	148	1785円
6.	ごみゼロ社会は実現できるか	行本　正雄 西本　哲 共著 立田　真文	142	1785円
7.	太陽の恵みバイオマス ― CO_2を出さないこれからのエネルギー ―	松村　幸彦著	近刊	

ヒューマンサイエンスシリーズ

(各巻B6判)

■監　修　早稲田大学人間総合研究センター

	書名	著者	頁	定価
1.	性を司る脳とホルモン	山内　兄人 新井　康允 編著	228	1785円
2.	定年のライフスタイル	浜口　晴彦 嵯峨座　晴夫 編著	218	1785円
3.	変容する人生 ―ライフコースにおける出会いと別れ―	大久保　孝治編著	190	1575円
4.	母性と父性の人間科学	根ケ山　光一編著	230	1785円
5.	ニューロシグナリングから 知識工学への展開	吉岡　亨 市川　一寿 編著 堀江　秀典	160	1470円
6.	エイジングと公共性	渋谷　厚 空閑　望樹 編著	230	1890円
7.	エイジングと日常生活	高木　知和 田戸　功 編著	184	1575円
8.	女と男の人間科学	山内　兄人編著	222	1785円
9.	人工臓器で幸せですか？	梅津　光生編著	158	1575円
10.	現代に生きる養生学 ―その歴史・方法・実践の手引き―	石井　康智編著	224	1890円
11.	いのちのバイオエシックス ―環境・こども・生死の決断―	木村　利人 掛江　直子 編著 河原　直人	224	1995円

定価は本体価格＋税5％です。
定価は変更されることがありますのでご了承下さい。

◆図書目録進呈◆

技術英語・学術論文書き方関連書籍

マスターしておきたい 技術英語の基本

Richard Cowell・余 錦華 共著
A5／190頁／定価2,520円／並製

本書は，従来の技術英語作文技法の成書とは違い，日本人が特に間違いやすい用語の使い方や構文，そして句読法の使い方を重要度の高い順に対比的に説明している。また理解度が確認できるように随所に練習問題を用意した。

科学英語の書き方とプレゼンテーション

日本機械学会 編／石田幸男 編著
A5／184頁／定価2,310円／並製

本書は情報化，国際化が進む現在，グローバルな技術競争の中で，研究者や技術者が科学英語を用いて行うプレゼンテーションや論文等の書類作成の方法を，基礎から実践まで具体的な例を用いてわかりやすく解説している。

いざ国際舞台へ！
理工系英語論文と口頭発表の実際

富山真知子・富山 健 共著
A5／176頁／定価2,310円／並製

ルールを知れば英語で研究論文を国際舞台に送り出してやることは，そう困難なことではない。本書は英語という言語文化にのっとった書き方，発表の仕方をまず紹介し，その具体的方法やスキル習得の方策を解説した。

知的な科学・技術文章の書き方
－実験リポート作成から学術論文構築まで－

中島利勝・塚本真也 共著
A5／244頁／定価1,995円／並製

日本工学教育協会賞（著作賞）受賞

理工系学生と若手の研究者・技術者を対象に，実験リポートと卒業論文のまとめ方，図表の描き方，プレゼンテーション原稿の作成法，校閲者への回答文の執筆要領，学術論文の構築手順などすべての科学・技術文章の書き方を知的に解説。

知的な科学・技術文章の徹底演習

塚本真也 著
A5／206頁／定価1,890円／並製

工学教育賞（日本工学教育協会）受賞

本書は「知的な科学・技術文章の書き方」に準拠した演習問題集である。実験リポート，卒業論文，学術論文，技術報告書を書くための文章と図表作成に関して徹底的に演習できる。文部科学省特色GP採択，日本工学教育協会賞を受賞。

定価は本体価格+税5％です。
定価は変更されることがありますのでご了承下さい。

◆図書目録進呈◆